The OpenMP Common Core

**Scientific and Engineering Computation**
William Gropp and Ewing Lusk, editors; Janusz Kowalik, founding editor

A complete list of books published in the Scientific and Engineering Computation series appears at the back of this book.

# The OpenMP Common Core

Making OpenMP Simple Again

Timothy G. Mattson, Yun (Helen) He, and Alice E. Koniges

The MIT Press
Cambridge, Massachusetts
London, England

This book was set in LaTeX by the authors and was printed and bound in the United States of America.

Library of Congress Cataloging-in-Publication Data

Mattson, Timothy G., 1958-
The OpenMP common core : making OpenMP simple again / Timothy G. Mattson, Yun (Helen) He, and Alice E. Koniges.
Cambridge, Massachusetts : The MIT Press, 2019. — Series: Scientific and engineering computation — Includes bibliographical references and index.
LCCN 2019033350 — ISBN 9780262538862 (paperback) Subjects: LCSH: Parallel programming (Computer science) — Application program interfaces (Computer software) — OpenMP (Application program interface)
LCC QA76.642 .M379 2019 — DDC 005.2/75–dc23
LC record available at https://lccn.loc.gov/2019033350

10 9 8 7 6 5 4 3 2 1

Dedicated to our families and our precious kayaks.

# Contents

## II      The OpenMP Common Core

# Series Foreword

The Scientific and Engineering Computation Series from MIT Press aims to provide practical and immediately usable information to scientists and engineers engaged at the leading edge of modern computing. Aspects of modern computing first presented in research papers and at computer science conferences are presented here with the intention of accelerating the adoption and impact of these ideas in scientific and engineering applications. Such aspects include parallelism, language design and implementation, systems software, numerical libraries, and scientific visualization.

William Gropp and Ewing Lusk, Editors

# Foreword

I am proud of OpenMP. It's been over 20 years since I joined a small group of parallel programmers to create OpenMP; and today, the language is still going strong. It is actively used by untold thousands of programmers and shows no sign of slowing down.

However, all is not well in the world of OpenMP. It is in danger of choking on the creeping complexity that so often plagues successful programming languages. New hardware emerges and in response, new features are added to OpenMP. Corner cases appear in a small number of algorithms that are difficult for OpenMP to handle, so new clauses are added to familiar constructs. Algorithms not considered when we created OpenMP are deemed critical and new constructs are defined. After 20 years of this process, it's hard to recognize the simple API we envisioned when this journey began way back in 1996.

With this book, we return to the simplicity we had in mind when we created OpenMP. The complexity was added for a reason. We are not trying to turn back the clock on the advances incorporated into OpenMP. Instead, we want to change how we teach OpenMP. People new to OpenMP should learn the small set of items from OpenMP that are used in most programs most of the time. Once this set of common, core constructs are mastered, then and only then should the more nuanced and complex parts of the language be explored.

If you are new to OpenMP, this is the book for you. Join us and learn the Common Core of OpenMP. With that foundation to build on, the sky is the limit as you explore a wide range of parallel algorithms on modern hardware.

Tim Mattson, May 2019

# Preface

This book is not a reference guide to OpenMP. You can find reference guides to the language on the OpenMP web site. That coupled with the text of the OpenMP specification provides the information you need. Also you can use the excellent book *Using OpenMP - The Next Step* by Ruud van der Pas, Eric Stotzer, and Christian Terboven [13] as a reference guide.

This book is about how to learn OpenMP. We assume our readers have no experience with multithreading and no knowledge of OpenMP. We introduce material in discrete chunks ordered to support effective learning. This is different from a reference guide where you go through the key elements of a system describing in full each element one by one. In this book, we introduce a few ideas and the OpenMP constructs supporting those ideas. Then later, as we move into more complex ideas, we revisit OpenMP constructs and describe additional aspects of that construct. For any given OpenMP construct, the full description of everything you can do with it may be spread out across several chapters.

This would be insane in a reference guide. But as you will hopefully see, this is great for learning a new programming language. For example, when teaching a child the concept of a mathematical function, you would never introduce the concept of limits and the value of the function as you approach a singularity. You would wait, often many years, as the child mastered the concepts of functions and then, after mastery is achieved by long practice, finish the full definition of functions by including limits. Likewise for OpenMP. Explaining all the mechanisms for controlling the data environment when introducing the parallel construct for creating threads would be overwhelming. It's much better to introduce thread creation and how to do useful work with threads. Then later, after the basics of managing threads has been mastered, we return to thread creation but this time we add the ability to manipulate the data environment.

The key to using this book is to actively work through the material. Download an OpenMP compiler (the gcc compilers support OpenMP as do most commercial compilers). As each OpenMP directive or API routine is introduced, write programs to experiment with them. Use them in different ways and understand how they work before moving on in the text. Do not just read the book cover to cover before writing code. Pause and write code as you read the book.

To support this style of *active learning*, our web site (`http://www.ompcore.com`) includes a wide assortment of programs and exercises for you to work with. Consult that web site often. We intend to update it often as we learn more during this continuing journey through the *Common Core* of OpenMP.

We close with a word about programming languages. OpenMP supports C, C++, and Fortran. Ideally, this book would include examples in all three languages. To do so, however, would greatly expand the size and scope of this book; and for all that extra work, there really would be little advantage to our readers. OpenMP is basically the same between the three languages. With very few exceptions (all of which we spell out in the book) if you know OpenMP for one language, you know it for all three. Therefore, we chose to define the constructs for C/C++ and Fortran, but the examples in the book and the bulk of the discussion centers on C. We believe this was the right compromise since C is the lowest common denominator among programmers in High Performance Computing. Even programmers who mostly write Fortran, know the basics of C.

To help our Fortran readers, we provide Fortran versions of all our examples on our website (`http://www.ompcore.com`). For those few Fortran programmers who do not know C, we also provide a short tutorial on the C programming language. We believe our book, when combined with these online resources, will be a great learning resource for Fortran programmers. Therefore, please do not let all the C code turn you away from our book. If you want to learn OpenMP, this book will help you regardless of whether you code in C, C++, or Fortran.

## Acknowledgments

The content of this book was painstakingly developed over the course of 20 years of teaching OpenMP. The examples, the basic flow of the material, the ways concepts are described ... all of this was worked out by teams of instructors during tutorials at various Super Computing Conferences. We'd especially like to thank Mark Bull (EPCC), Sanjiv Shah (Intel), Barbara Chapman (Stony Brook University), Larry Meadows (Intel), Paul Petersen (Intel), and Simon McIntosh-Smith (Bristol University). The memory model content was particularly difficult to pull together. Xinmin Tian (Intel), Michael Klemm (Intel) and especially Deepak Eachempati (Cray) were instrumental in helping us define that material. To develop the memory model examples, we needed access to a wide variety of architectures. Simon McIntosh-Smith was most helpful by giving us access to the Isambard system at the University of Bristol.

We greatly appreciate the review team pulled together by MIT Press. Their feedback helped us tremendously as we moved from *final draft* to *final copy*. Among this group, we single out Ruud van der Pas (Oracle). Ruud is a good friend who has encouraged us in this project from the beginning. He went above and beyond the call of duty with an unusually thorough review of the book.

We would also like to thank the OpenMP Architecture Review Board for letting us use excerpts from the OpenMP Specification and Examples documents. Finally, we want to thank people from across the OpenMP community: Bronis de Supinski (LLNL) and the OpenMP Language Committee he leads, OpenMP programmers we have worked with, and all the students who have taken our tutorials in the past. We could not create a book like this without you.

# The OpenMP Common Core

# I SETTING THE STAGE

From the early days of computing and even more so today, performance comes from doing many things at once. With the pipelined execution units in the original Cray vector supercomputers, distributed memory workstation clusters, and many cores in a single CPU; parallelism has been the key to performance for decades.

Parallel hardware is only useful with parallel software. It would be nice if parallel software could be generated automatically. This has been tried many times and with very few exceptions, does not work. Programmers need to write parallel software.

This has created a tension between the hardware and software communities. New hardware emerges and programmers must adapt. This means the programming languages and tools must adapt along with the hardware.

In this first part of the book, we will describe the world within which parallel programmers operate. We will set the stage and focus on the needs of programmers working with one important class of parallel systems: shared memory multiprocessor computers. This discussion will guide us to the programming models used for these machines and the historical roots of OpenMP.

# 1 Parallel Computing

A program is a stream of instructions to run on a computer. A computer is made up of many parts including memory, a storage system, and one or more processing elements. A processing element is the part of the computer that actually does the "computing" and executes instructions.

A program is called a *sequential program* when it runs on a single processing element. A parallel program runs multiple streams of instructions at the same time on multiple processing elements. The basic idea is simple, but using parallelism to solve real problems is anything but simple.

Parallel computing has grown into its own distinct branch of computer science. It has its own specialized jargon, its own set of concepts, its own hardware, and of course, its own programming languages. It developed within the high performance computing (HPC) community and today is part of the core knowledge assumed of any HPC programmer. For the wider community of programmers outside of HPC, however, many of the ideas in parallel computing are foreign.

We want everyone to be able to read and benefit from this book, not just HPC programmers. Therefore, in this chapter we develop the language and conceptual foundations of parallel computing. For experienced HPC programmers, you can quickly skim this chapter to verify that you use the specialized jargon of parallel computing the same way we do. For people new to HPC or from outside the HPC community, you will want to read this chapter carefully as this is where we establish the conceptual foundation you will need to learn OpenMP.

## 1.1 Fundamental Concepts of Parallel Computing

If you write programs, you are probably familiar with the basic structure of a computer. There is an address space that holds both data and the text of your program. We call this the memory of the computer. The term "data" refers to a set of variables that name addresses in memory and reference a set of values stored at those addresses. The computer's control unit loads instructions from the program stored in memory and carries out the operations from this single stream of instructions to produce results. This simple concept of a sequential computer, later known as the von Neumann architecture, dates back to a 1945 report on the design of early computers by John von Neumann [14].

All programmers learn to write programs for sequential computers. Initially this was enough. The central processing unit (CPU) at the core of a computer was able to deliver ever-increasing performance gains demanded by the market. The CPU did this by becoming increasingly more complex to deal with the various bottlenecks that limited performance.

Let's consider the CPU in more detail. It consists of an *arithmetic logic unit* (ALU) which performs arithmetic and logical operations plus a control unit that manages the flow of data and instructions. Modern CPUs extract parallelism at the instruction level by breaking down instructions into smaller micro-ops which are fed into a processing pipeline. The control unit keeps track of dependencies between micro-ops so they can execute in parallel or even out-of-order while still delivering the same results as you would expect from the original serial stream of instructions. This results in a form of parallelism known as superscalar execution. Fortunately, superscalar execution is managed by the CPU on behalf of the programmer who continues to think in terms of a single, sequential stream of instructions. With the

---

**The Harsh Reality of Hardware Trends**: Moore's law states that every 2 years or so, the number of transistors on a semiconductor device (or "chip") will double. It is a prediction of an economic trend, not a law of physics. Moore's law dates back to 1965, and up until 2004, it resulted in ever faster chips. Features etched onto a chip became smaller and smaller and the energy required to switch states of transistors (the dynamic energy) went down. This is called Dennard scaling [3], which noted that with smaller features on a chip, the voltage decreased and chips could be driven at higher frequencies.

Around 2004, Moore's law stopped delivering higher clock speeds. The dynamic energy for switching transistors continued to decrease. Leakage and other static energy demands did not decrease. Eventually, they came to dominate the energy needed to drive a chip and Dennard scaling ended. Moore's law continued to shrink transistors. With the end of Dennard scaling, however, the benefits from increasing transistor counts could only arise from architectural innovation. That means more cores, wider vector units, special purpose accelerators, etc.

From a hardware point of view, this is great. Hardware engineers have much more fun in the post-Dennard-scaling era. It is the software community that suffers. Hardware engineers throw increasingly complex *stuff* at software developers and programmers who care about performance have no choice; they must deal with this harsh hardware reality and figure out how to write programs for these parallel, heterogeneous devices.

rise of commercial off-the-shelf CPUs for the mass market and the economic engine of Moore's law, transistor densities doubled every two years and the performance followed suit. There was little reason, except for the most aggressive supercomputing applications, to worry about moving beyond the sequential model of computing.

This all changed around 2004. As explained in the boxed text, software developers who care about performance have no choice but to write parallel code. Parallel programming is not optional and should be part of every software professional's skill set.

## 1.2 The Rise of Concurrency

To understand parallel computing, we have to start with the closely related concept of *concurrency*. Two or more streams of instructions are said to be *concurrent* if the individual instructions from any one stream are unordered with respect to the instructions from the other streams [7].

This is best understood with a simple example. Using your favorite editor, type in the code in Figure 1.1. Do not worry about the meaning of the pragma in this code. We will describe that later.

```
1   #include <stdio.h>
2   #include <omp.h>        // The OpenMP include file
3
4   int main()
5   {
6       #pragma omp parallel
7       {
8           printf(" Hello ");
9           printf(" World \n");
10      }
11  }
```

Figure 1.1: **A simple "Hello World" C program to demonstrate concurrent execution**.

Compile this code using an OpenMP enabled compiler such as GCC. To enable the compiler to recognize OpenMP directives, you must set a (compiler dependent) flag. For example, with GCC, you must use the -fopenmp option to the compiler to tell it to create a multithreaded program using OpenMP. Then run this program as you would any other executable.

```
$ gcc -fopenmp hello.c
$ ./a.out
```

The system will give you the default number of threads which if you are running on a typical laptop today will be around 4 threads (the number of cores seen by the operating system). With a serial programming mindset, you might expect the output from the program to be:

```
Hello World
Hello World
Hello World
Hello World
```

The four threads, however, execute concurrently. The instructions executed by each thread are unordered with respect to the other threads. The printf statements within each thread follow the order defined by the program, but between the different threads, they have no specified order at all. So the output may be something like:

```
Hello Hello World
World
Hello   World
Hello   World
```

Furthermore, each time you run the program, the output may be different. Since the threads are concurrent, each time the operating system schedules the threads for execution, the order of the output operations may change. Another way to think about this is that every legally allowed way to interleave the statements defines a possible execution order for the program. Your challenge as a parallel programmer is to make sure that all possible interleavings yield correct results.

You have now run your first multithreaded program. You see that the available threads execute concurrently. Let's bring the concept of parallelism or "parallel execution" into the conversation. If a set of concurrent threads execute on distinct processing elements so they make progress at the same time, they are said to execute in parallel. Concurrency defines operations that can execute in any order (i.e., they are unordered). Parallelism brings multiple hardware elements into the conversation so operations run at the same time. Note that concurrent and parallel have different meanings, although occasionally you may notice these terms are confused to mean the same thing. That is not correct. Parallel execution via hardware allows for the simultaneous execution of concurrent tasks.

## 1.3 Parallel Hardware

With the end of Dennard scaling, the focus in computer design shifted from an emphasis on ever increasing clock speeds to clever architectural features that exploit parallelism. This led to a wide range of systems: distributed memory clusters, programmable GPUs, vector units that drive multiple data elements from a single stream of instructions (SIMD or "Single Instruction Multiple Data") and multi-processor computers. OpenMP is effective for all of these other than the distributed memory clusters.

### 1.3.1 Multiprocessor Systems

OpenMP started with a focus on multiprocessor systems. For most OpenMP programmers, these systems are still their primary concern when writing OpenMP programs. A multiprocessor computer consists of multiple processors that share an address space. The memory available to the processors is shared; hence why they are often referred to as shared memory computers. To understand a multiprocessor system, we use a basic model that makes the essential elements of the system explicit while hiding details deemed less essential. The model we start with for multiprocessor systems is the Symmetric Multiprocessor or SMP model shown in Figure 1.2.

An SMP has N processors that share a single memory. The hardware is managed by an operating system (OS) which treats all the processors the same. Furthermore, the cost of accessing any variable in memory is the same for any processor. In other words, the processors in this model are "symmetric" from the perspective of the OS and memory.

The SMP model is a gross oversimplification. It is unlikely that you will ever encounter a computer that looks like an SMP system. Let's consider a modern processor in more detail. A CPU is a general purpose processor that sits in a single socket within the computer. It is optimized to deliver results from individual

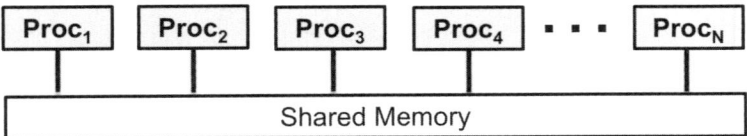

Figure 1.2: **A Symmetric Multiprocessor computer consisting of N Processors (Procs) sharing a region of memory.**

events quickly, that is, the CPU is optimized for low latency. A modern CPU is a multiprocessor computer. Multiple distinct processors called *cores* are placed on a single silicon die packaged into a multicore CPU. We show a schematic representation of such a multicore CPU in Figure 1.3.

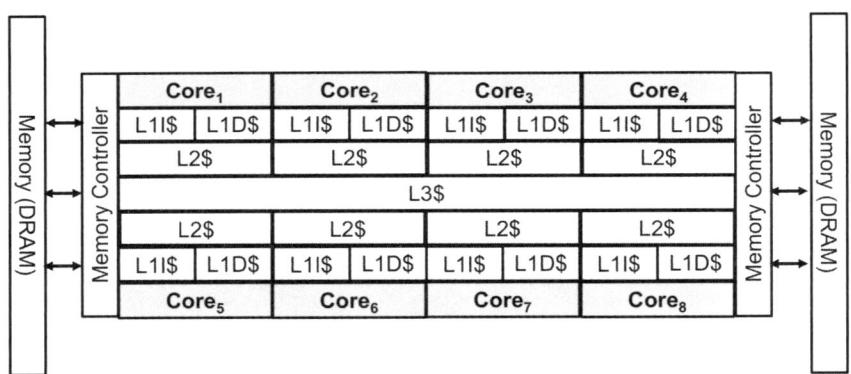

Figure 1.3: **A typical multicore CPU with 8 cores** – Each Core has level one caches for data (L1D$) and instructions (L1I$), a unified level 2 cache (L2$) and a shared level 3 cache (L3$). This CPU has two memory controllers each with three channels to access off-chip memory (DRAM).

The memory for a multicore system is far more complicated than that for an SMP system. All processors share the memory which is exposed to them as a single address space. The DRAM (Dynamic Random Access Memory) used to implement memory in a computer is much slower than the cores within a chip. Therefore, modern CPUs include small regions of high speed memory tightly integrated into the cores themselves called caches.

A cache is not accessed as a separate address space. Memory within the cache is mapped onto the addresses in DRAM. The details of this mapping are well beyond the scope of this book. For our purposes, think of the cache as a high speed window into the larger memory in DRAM. The mapping between the cache and DRAM is carried out in small blocks called a cache line. A typical cache line is just 64 bytes (i.e., enough space to hold 8 double-precision floating-point numbers). The reason caches are organized into cache lines is that in many programs, if an instruction needs a value at an address $N$, it is likely the next instruction will access the value at address $N + 1$. By exploiting this *spatial locality*, performance can be greatly enhanced.

Memory is hierarchical in a modern system. Consider a typical multicore CPU as shown in Figure 1.3. There is a small cache right next to each processor to hold data and instructions for the program. These are called level one caches since they are the closest to the cores; the L1D\$ (Data) and L1I\$ (Instruction) caches. Each core has a level 2 cache (L2\$) that holds both data and program instructions, hence this is called a unified cache. Finally, all the cores in a CPU may share an additional level of cache, a level 3 or L3\$. The caches are small. For a typical high-end CPU, the sizes[1] of these caches are:

- 32 KiloBytes Level 1 Data cache per core

- 32 KiloBytes Level 1 Instruction cache per core

- 256 KiloBytes Level 2 cache, unified (holds both data and instructions) per core

- 8 MegaBytes Level 3 cache shared between cores

Cache lines are constantly moving through the memory hierarchy as a program executes. Multiple cores may at any given time access any of the cache lines creating room for memory access conflicts. A cache coherency protocol built into the CPU manages all these cache lines to assure that eventually all the cores will see the same values in memory. The key word here is "eventually" and it is possible that at any given moment, the cores may see different values for any given address. The set of rules that governs the values the different cores may see when accessing shared

---

[1]A byte is 8 binary values or bits. $2^{10}$ bytes is a *KiloByte* or 1024 bits. A MegaByte is 1024 KiloBytes or 1048576 bits.

addresses in memory is called the *memory model*. We will discuss the memory model in detail later in the book.

A key aspect of the SMP model is that every processor has an equal cost to any address in memory. Even a cursory glance at the multicore chip in Figure 1.3 suggests that this is not the case. The time required to access a value in memory depends on where that value is in the memory hierarchy. Consider the time required to access a single value from memory, that is, the latency of a memory access. Using values for a typical, high-end CPU the values range across the memory hierarchy are as follows:

- L1 Cache latency = 4 cycles

- L2 Cache latency = 12 cycles

- L3 Cache latency = 42 cycles

- DRAM access = $\tilde{2}50$ cycles

Therefore, rather than a memory system with uniform memory access times, the reality is that systems have a nonuniform memory architecture (NUMA); that is, we say that modern multiprocessor systems are not SMPs, they are NUMA systems. It is even worse than that implied by the multicore CPU in Figure 1.3. High end servers, especially those used in HPC systems, often connect several CPUs into a large NUMA cluster. For example, we show a typical block diagram for a server used in HPC systems with four CPUs connected by high speed point-to-point processor interconnects in Figure 1.4. Each CPU in the cluster has its own memory controller and its own block of DRAM. All the memory in the 4-CPU cluster is organized into a single (shared) address space and is accessible by any core in the system. Without going into detailed timings, it is easy to see that the cost of accessing memory will vary widely as cores access cache lines mapped to their own chip's DRAM or a DRAM bank associated with some other chips in the system.

Parallel hardware can be very confusing. Fortunately, with cache coherence protocols moving cache lines around as needed and modern optimizing compilers, you can get away with thinking in terms of the SMP model as you write your code. Then as a later optimization step, you modify your program to take advantage of the NUMA features of your individual system. This includes straightforward optimizations such as reorganizing loops to improve reuse of data from cache lines (cache blocking) as well as more complicated optimizations such as initializing data

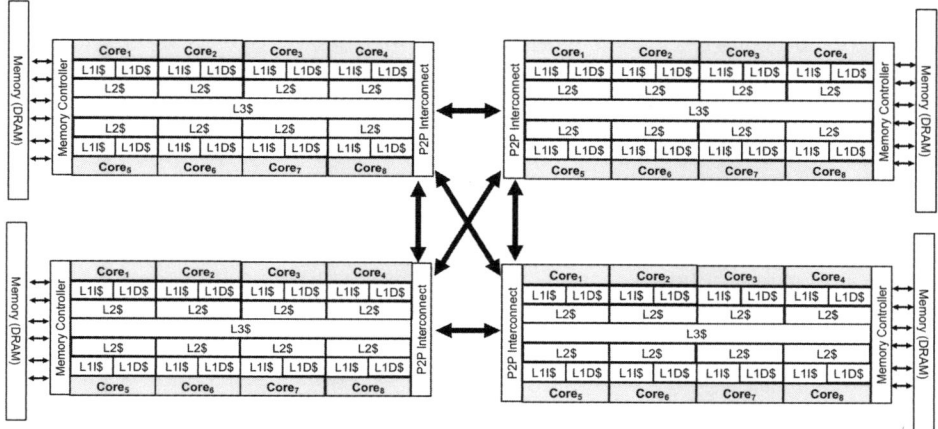

Figure 1.4: **A nonuniform memory architecture (NUMA) system with four CPUs connected by point to point (P2P) interconnects** – All DRAM is accessible to all the cores meaning the cost of accessing different regions of memory varies significantly across the system.

on the same cores that will later process that data. We discuss these and other related optimizations later in this book.

### 1.3.2 Graphics Processing Units (GPU)

The central problem in computer graphics is to take a model for a scene and turn it into pixels for presentation to a (usually human) visual processing system. Using a mix of computer science, physics, lots of mathematics, and a good dose of psychology, the resulting images convey meaning to the viewer. The processing carried out in graphics is naturally data parallel: a scene is broken down into geometric figures which are further turned into a collection of fragments. The fragments stream through a processing pipeline and are rendered into a set of pixels displayed on a screen. We are simplifying the process significantly, but overall the processing pipeline consists of phases like these:

- **foreach 3D object** tile the visible surface with polygons.

- **foreach polygon** decompose into triangles.

- **foreach triangle** compute the color **foreach** pixel.

The key idea here is **foreach** followed by the name of a collection of some sort (e.g., the set of polygons or the set of triangles). The function (e.g., decompose into triangles) is applied to each member of the collection (e.g., the set of polygons). The computations on each member of the set are independent and can occur at the same time; that is, the parallelism is in the data (hence, the name *data parallelism*).

Specialized hardware to support some version of a graphics pipeline goes back to the 1970s. Over time, the processing carried out **foreach** member of a set of objects became more complicated. Rather than fixing that function in the hardware, it became important to be able to program those units. Bit by bit the hardware used for graphics became less "hardware specialized" and more and more programmable.

In 2006, fully programmable GPUs showed up on the market. Proprietary Application Program Interfaces (APIs) were available from GPU vendors who talked about GPGPU or General Purpose GPU programming. Developments moved fast in the GPGPU world. Two years after GPGPU devices first showed up on the market, a standard programming language for GPGPU programming called OpenCL [11] was released. These standard-based solutions (OpenCL and OpenMP) and proprietary solutions have continued to evolve rapidly making GPGPU programming a mainstream skill for performance-oriented programmers.

Clearly, a detailed discussion of GPUs and how to program them is way beyond the scope of a book about learning OpenMP. We want to describe the key features for GPGPU programming, however, to leave you with a well-rounded exposure to the foundations of parallel programming. Then much later in Section 12.3, we will revisit this topic and discuss how to do GPGPU programming using OpenMP.

The most fundamental issue that divides GPGPU from CPU programming is throughput *vs.* latency. When a stream of frames is displayed on your monitor, you do not care how long it takes to update a single pixel. All you care about is the rate whole frames are produced; that is, are they produced at a fast enough rate so your videos look smooth. In other words, you care about the *throughput* of the frames moving across your screen. Contrast that with the response to a command submitted to your computer. If you click your mouse at a point on your screen, you expect an immediate response. You care about the *latency* of an individual computation.

A great amount of circuitry is invested to maintain the low latencies demands of a general purpose processor. For example, CPUs include cache hierarchies dedicated to keeping the memory close to the processing elements that will work with it. A GPU is optimized for throughput. Transistors are organized to pack many processing elements onto a single chip. The memory system for the chip supports high throughput but it lacks the complex multilevel caches that CPUs use to support low latencies.

The basic idea of the throughput optimized GPU programming model is to turn the index range from the foreach statements into an index space. The body of the foreach statements is turned into functions called *kernels*. An instance of the kernel is run at each point in the index space (an *NDRange* or *grid*). We call this kernel instance a *work-item*. Data is aligned to the same NDRange so we keep data close to the work-items that the processing elements will manipulate it.

To make a GPU a specialized throughput engine, the scheduler built into a GPU knows about the data needed by a kernel. The work-items are grouped together into *work-groups* and a work-group waits to execute until its data is available. If you have many more work-groups than processing elements, they queue up and there is plenty of work available to keep the processing elements busy while waiting for data to stream through the system. This is an example of *latency hiding*; that is, overlapping memory movement with computation so you do not see the high cost of accessing individual blocks of memory.

When coupled with a memory system optimized for high bandwidth, GPUs can deliver high throughput for data-parallel workloads. They do not handle all workloads. If work-items need to interact, the programming becomes more complicated and at some point, the approach based on hiding memory latencies with deep queues of work-groups falls apart. With a great deal of creativity, a wide range of algorithms have been mapped onto the GPGPU data parallel model of computation. And as we will see in Section 12.3, you can do all of this in OpenMP.

### 1.3.3   Distributed Memory Clusters

In scientific computing, given a problem of size $N$, someone working in that scientific domain will most likely want to run a problem of size $10 \times N$ or larger. At the upper ends of scientific inquiry, there seems to be an insatiable need for more and more computing. This means that individual processors, whether they are CPUs or GPUs, cannot always keep up with demand.

It is possible to integrate multiple processors into a single memory domain. We did this, for example, with the NUMA system in Figure 1.4. Creating a NUMA system takes specialized hardware engineering and is very expensive. As the hunger for more and more computing increases, NUMA systems won't be able to keep up either.

The solution is to limit how far you push NUMA systems. Instead, take standard off-the-shelf servers built for massive data centers and network them together into a large system. Then use software to tie them together into large parallel computers. A computer like this, built from off-the-shelf components, is called a *cluster*. As long as you can pay the electric bill to run them and support a large enough building to hold them, a cluster can be as large as you want and support truly incredible rates of computation.

As a parallel computer, a cluster is quite different from the shared memory systems we have considered so far. The servers in the cluster sit at nodes in the cluster's network. When nodes are specialized for dense compute, perhaps by pairing a GPU with a high-end CPU or by combining multiple CPUs into a NUMA domain, they are sometimes called *compute nodes* to distinguish from the nodes optimized for more typical data-center operations (sometimes called *server nodes*). The nodes do not share physical memory, so the memory is distributed around the system. The nodes interact by passing messages to each other. The key idea is "two-sided" communication where one node sends a message to another node that posts a command to receive that message. The standard API for programming distributed memory machines (such as clusters) is called the Message Passing Interface or *MPI*. There are higher level systems for programming distributed memory machines based on one-side communication, partitioned global address spaces (PGAS), and map-reduce frameworks. For HPC, however, the ubiquitous common denominator is MPI.

We will not cover MPI in this book. If you work in HPC, you will need to learn MPI at some point. The dominant model in HPC is MPI between nodes and OpenMP within a node. We call this the hybrid MPI/OpenMP model. Even if you use high level PGAS languages or some of the newer task driven programming models, underneath them sits MPI and, if not OpenMP, a multithreading model similar to OpenMP.

## 1.4   Parallel Software for Multiprocessor Computers

Multiprocessor computers have been around for over 50 years. We have figured out how to organize software to manage these systems. This greatly simplifies life for a programmer since one common view of how software is organized for these systems is likely to work for any system you encounter.

An operating system manages the hardware on behalf of the user of the system (which includes the programmer). When you launch a program, the operating system creates a process. Associated with this process is a region of memory and access to system resources (e.g., the file system). A process forks one or more threads which are all part of a single process. Each thread has its own block of memory, but all the threads share the memory of the process as well as the system resources available to the process. The threads carry out the instructions of the program on behalf of the process.

The operating system schedules the threads for execution. There are many more threads than processors within a system. The basic idea is for the operating system to swap concurrent threads in and out of execution. This way if a thread is blocked while waiting for some high latency event (such as a file access), other threads that are ready to execute can swap in and exploit available processors. The goal is to efficiently exploit the processors on the multiprocessor computer so none of the processors are idle for a significant fraction of time.

An Application Programming Interface (API) is an interface to functions, data types and any other building blocks a programmer can use when writing software. The idea is to hide complex and potentially dynamic features of a system behind a fixed API. The operating system provides a low-level API to manage threads. Most operating systems, including those based on Linux (which include Apple's OSX), support the IEEE POSIX threading model called *pthreads*. As with any low level API, using pthreads directly gives you lower overheads and maximum flexibility when working with threads. Writing code directly with the POSIX pthreads API, however, is tedious, error-prone, and not something most application programmers would be willing to do. For example, in Figure 1.5 we show the pthreads version of the "Hello World" program we discussed earlier in Figure 1.1. It is not our goal to teach you how to write code with pthreads. We just want to highlight some of the high level features of pthreads programing and emphasize that for most programmers, you *DO NOT* want to write code this way.

The first step in pthreads programming is to isolate the code you want to execute in a thread as a function. In this case the function is `Print_HelloWorld(void *)`. Inside the main function, we setup an array to hold thread IDs and a pthreads opaque object to hold the attributes of the thread. We then run a loop over the number of threads we wish to create and create (i.e., "fork") the threads, passing a

```
1   #include <pthread.h>
2   #include <stdio.h>
3   #include <stdlib.h>
4   #define NUM_THREADS 4
5
6   void *PrintHelloWorld(void *InputArg)
7   {
8       printf(" Hello ");
9       printf(" World \n");
10  }
11
12  int main()
13  {
14      pthread_t threads[NUM_THREADS];
15      int id;
16      pthread_attr_t attr;
17      pthread_attr_init(&attr);
18      pthread_attr_setdetachstate(&attr, PTHREAD_CREATE_JOINABLE);
19
20      for (id = 0; id < NUM_THREADS; id++) {
21          pthread_create(&threads[id], &attr, PrintHelloWorld, NULL);
22      }
23
24      for (id = 0; id < NUM_THREADS; id++){
25          pthread_join(threads[id], NULL);
26      }
27
28      pthread_attr_destroy(&attr);
29      pthread_exit(NULL);
30  }
```

Figure 1.5: **A simple "Hello World" C program using pthreads.**

pointer to the function each thread should run. Later, the program waits for each thread to complete at which point the thread exits; in essence, the execution joins with the main flow of the program and once all threads complete their join, the main program proceeds to clean up the pthreads objects and terminate.

This is one of the simplest examples of pthreads programming. The code gets much more complicated when you pass arguments to threads, control the threads with additional attributes, and order memory operations between threads. When you need complete control over the threads to maximize performance or manage all the ways the threads can interact with the system, there may be no choice but to code at the level of pthreads. Most programmers, however, will not tolerate programming at such a low level. For most programmers, a higher level of abstraction is required. This is a primary reason that the OpenMP standard was introduced.

For programmers interested in building applications rather than operating systems or low level services, coding takes place in terms of a high-level model. The key is to pick a model well suited to the target hardware. Remember that hardware is replaced every few years while a good application program will last for decades. Hence, when choosing a high level model, it is important to select one that is not tied to hardware from a single vendor. A standard programming model supported by all relevant vendors is essential. For example, MPI, pthreads, and OpenMP are available from multiple sources (including open source implementations) and run on all mainstream platforms.

The programming model must also be well matched to the sorts of algorithms or the fundamental design patterns exploited in your applications. For example, OpenMP is an excellent match for programs organized around nested loop structures and task level programs exploiting shared memory. It would be a poor choice for distributed-memory architectures and applications with hard real-time constraints.

Finally, a programming model is defined by a specification. You cannot compile a program with a specification. You need a programming environment, or in the case of OpenMP, compilers that support a specification. It is common, and admittedly frustrating, for programming environments to take many years to fully support a new version of a programming model specification on a variety of architectures. OpenMP, due to heavy involvement from key vendors, does a good job of tracking new versions of the standard. In most cases, fully conformant implementations are available within a year of a new release of an OpenMP specification.

# 2 The Language of Performance

There are only two reasons to write a parallel program: to solve a fixed size problem in less time, or to solve a larger problem in a reasonable amount of time. In either case, it is all about performance. OpenMP is a programming language for writing parallel programs. At some level, it always comes back to performance.

Performance is such a simple term. However, the word hides layers of complexity and takes on different meanings depending on the context. Raw measures of performance are grounded in time, but even the seemingly unambiguous concept of "time" is nuanced with differences such as "CPU time" (the CPU frequency times the number of cycles expended by a CPU when actively executing your program) *vs.* "wallclock time" (the time as measured by a clock external to the computer, i.e., a clock "on the wall"). Performance is rarely interesting as a stand-alone number. We are usually concerned with performance as a comparative measure to highlight performance trends.

This complexity has led to a rich language of performance. Parallel programmers talk about speedup, efficiency, scalability, parallel overhead, load balancing, and a host of concepts used to reason about performance. The goal of this chapter is to explain this terminology.

## 2.1  The Basics: FLOPS, Speedup, and Parallel Efficiency

We write a parallel program to reduce our "time to solution". The time we care about is which we personally experience; that is, the wallclock time. As we convert a program into a parallel program, the wallclock time should decrease. With additional processors, the wallclock time should continue to decrease until the potential parallelism available from either the hardware or our algorithm is exhausted.

We can present timing data in ways that expose different aspects of performance. For example, if we express performance in terms of a rate; that is, the number of computations per second of wallclock time, we can immediately compare results to the peak performance of a computer or the theoretical performance we expect from our algorithms.

High performance computing (HPC) is largely centered on arithmetic with floating point numbers. Hence, we often think about performance in terms of floating point operations that can be completed in one second of wallclock time. This unit of measure is called FLOPS: floating point operations per second. We add to FLOPS the appropriate prefix, Mega ($10^6$), Giga ($10^9$), Tera ($10^{12}$), Peta ($10^{15}$) and so on to make the units convenient. Unless otherwise stated, the default floating point type when talking about FLOPS is a `double`, which on most computer systems is a 64-bit quantity. While HPC is focused on FLOPS, as you move away from the world of numerical simulations common to HPC into a broader range of computations, the focus sometimes shifts to *operations per second* (OPS) or *instructions per second* (IPS), but for most OpenMP programmers, it is almost always about FLOPS.

Let's put the range of FLOPS in context. In the late 1980s, a processing unit in the fastest supercomputers at that time (e.g., a Cray 2) ran with a peak performance of 500 MegaFLOPS (or 500 MFLOPS). A common iPhone today runs the Linpack[1] 1000 benchmark used to track performance in HPC systems at over 1200 MegaFLOPS. The Cori system at the National Energy Research Scientific Computing Center (NERSC), a Cray XC40 Supercomputer, ran the MPLinpack benchmark at 14 PetaFLOPS with 622,336 cores in 2018 (https://www.top500.org/system/178924). By the early 2020s, the world's fastest supercomputers should cross the ExaFLOP barrier ($10^{18}$ FLOPS).

Raw performance expressed in FLOPS is interesting, but to parallel programmers, it is not enough. We want to know how well our programs utilize the resources of a parallel system. We want to measure how much faster our programs run as we add processors. We want to know about the *speedup* of our programs.

Speedup is a ratio of run times for a program. Ideally you run a program with the best available serial algorithm on one processor. Then run it again with parallel software on $P$ processors. If we define $T_s$ as the run time for the serial program and $T_P$ as the run time for a parallel program running on P processors, we define the speedup as:

$$S(P) = \frac{T_s}{T_P}$$

it is not always possible to run an optimized serial program to measure $T_s$. In many cases, you implement a parallel algorithm and have no reason to implement the corresponding serial algorithm. In those cases, you replace $T_s$ with the time for the parallel program running on one processor, $T_1$. In other cases, the serial code may not fit in the memory of a single processor so we can only compare against a parallel program running on a minimal number of processors. With this potential

---

[1] The Linpack benchmark solves a dense system of linear equations: the famous $Ax = b$ problem. We used to run the benchmark for a matrix of fixed order (e.g., Linpack 1000), but now people run them for the largest problems that will fit on their machines, the MPLinpack benchmark.

ambiguity in the definition of speedup, it is important when reporting speedups to define the reference "serial execution" you are comparing against.

In the ideal case, the speedup is equal to the number of processors. If you double the number of processors, the performance should double. When a program follows this speedup trend we say that it has "perfect linear speedup". We measure how closely we track perfect linear speedup with a scaled speedup called the *parallel efficiency*.

$$eff = \frac{S(P)}{P}$$

We show the speedup and efficiency as a function of the number of processors (in this case, cores in a single node of an HPC cluster) in Figure 2.1. This program (octo-tiger: a program that uses a fast multipole algorithm for gravitational field simulation) uses a task driven system called HPX [5] and shows excellent speedups. Notice that for larger numbers of processors (as in Figure 2.1), we often use a logarithmic scale for both the horizontal and vertical axes.

Speedup is important for understanding a parallel algorithm and how it scales with the number of processors. You must never lose sight, though, that ultimately the goal is to reduce the time spent working on a problem. One of the oldest "cheats" in high performance computing is to pick a slow algorithm that happens to have a low serial fraction. It will show excellent speedup, but if the algorithm you start with is bad, the great speedup is really misleading you and masking the fundamental low quality of your solution. Hence, measure speedups and understand what they tell you about how your algorithms scale. Always take the time, however, to make sure you are using a good (i.e., fast if not optimal) algorithm that reduces the overall run time for your computation.

## 2.2   Amdahl's Law

As we have seen, a central concern in parallel computing is to understand how the performance scales with the number of processors. If we increase the number of processing elements, will the performance keep improving without bound? This question was addressed by Gene Amdahl in a 1967 paper [1]. This was a time when parallel computers were first appearing. His goal was to limit enthusiasm for the idea of building larger computers by combining many smaller processors. It led to a general principle we call Amdahl's law.

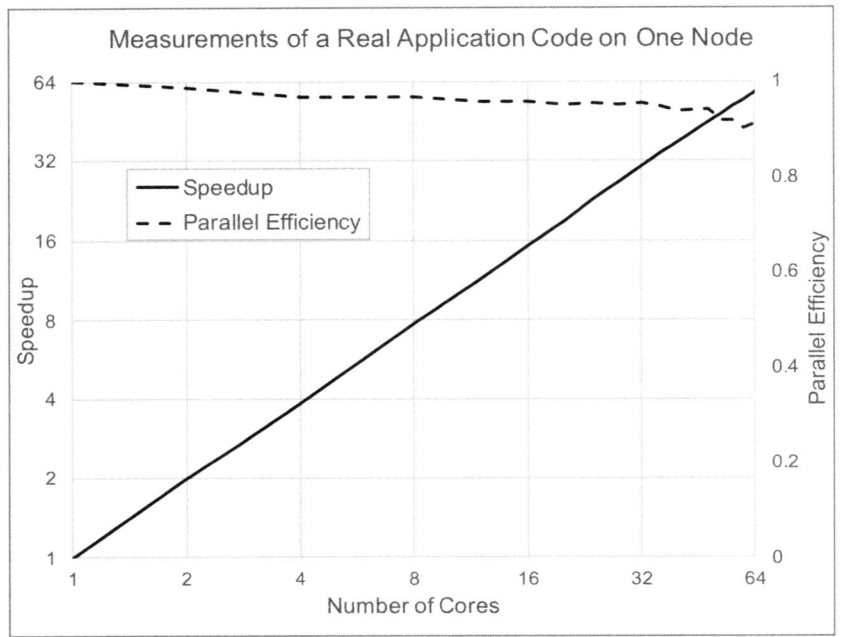

Figure 2.1: **Speedup and Efficiency as a function of the number of cores** – Performance measured on a single node of a cluster using log base 2 for the x and y axes for the speedup and log-linear scales for the corresponding parallel efficiency. The application is the fast multipole code Octo-tiger parallelized with the HPX task driven programming model. The node is an Intel(R) Xeon Phi(TM) 7250 68C processor running at 1.4GHz.

To derive Amdahl's law, we start with a basic simplification. Assume a program has some fraction of its work, $\alpha$, that is fundamentally serial. For that part of the program, it will not run faster as you add processing elements. We call $\alpha$ the *serial fraction*. If $\alpha$ is the serial fraction of a problem, then $(1 - \alpha)$ is the parallel fraction. Let the run time for the serial code be $T_s$ and the run time for the parallel code using $P$ processors be $T_p$. Within this model, the parallel run time is:

$$T_p = \alpha \times T_s + (1 - \alpha) \times \frac{T_s}{P}$$

with the time to execute the fraction of the code that is parallelizable reduced by a factor of $P$. The speedup is given by

$$Speedup = \frac{T_s}{T_p} = \frac{1}{\alpha + \frac{(1-\alpha)}{P}}$$

In the limit of $P$ approaching infinity, the parallel term approaches zero and we are left with Amdahl's law:

$$Speedup = \frac{1}{\alpha}$$

If your algorithm can speedup 95 percent of your program, for example, then your serial fraction is 0.05 and the best speedup you can achieve given unlimited number of processors is 20.

Consider the variation in the speedup as a function of $P$. In Figure 2.2, we show the maximum possible speedup for different values of the parallel fraction. According to Amdahl's law, if only 90% of the code can be parallelized, your serial fraction is 10% and you cannot do better than a speedup of 10 no matter how many processors you have to work with. The impact of the serial fraction of the code and the fact it cannot benefit from extra processors degrades the speedup well before we hit our limit at 10. As a good rule of thumb, good scaling requires that your serial fraction be an order of magnitude less than the limit suggested by Amdahl's law.

## 2.3   Parallel Overhead

Amdahl's law limits the benefits of parallelism due to limitations in the algorithm. If there are parts of an algorithm that cannot exploit multiple processors, then additional processors will not help. This is a serious limitation that parallel programmers must grapple with. Unfortunately, the problem is even worse than that suggested by Amdahl's law. In addition to the impact from the serial fraction of your algorithm, there is the problem of parallel overhead.

The parallel overhead is time spent to manage the threads in your parallel application. For example, later in this book, we will learn a great deal about the ways threads coordinate their execution. They are created, destroyed, and sometimes they wait for other threads to complete an action before proceeding. This adds wasted time to the run time of a program; or we say it adds overhead to the program execution. In such cases, this overhead grows with the number of threads.

We have mentioned the overhead of managing threads. Other sources of overhead arise form managing the data. If data must be distributed between processors, as

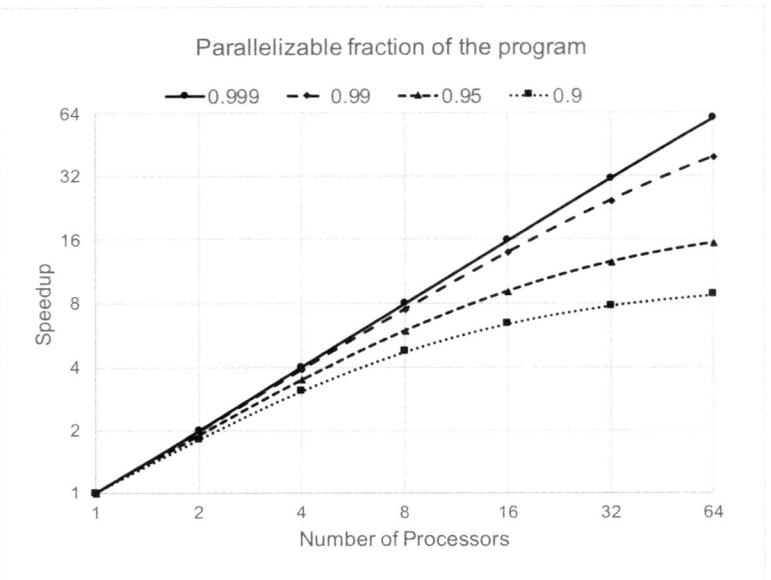

Figure 2.2: **Speedup as a function of the number of processors for different values of the parallelizable fraction of a program** – We plot the log of the Speedup *vs.* the log of the number of processors. Note that the speedup falls off quickly as the fraction of the program that can be parallelized drops from 0.999 to 0.9.

the number of processors grows the data movement between them grows. Memory speeds and network speeds are much less than processor speeds, so data movement can quickly grow to swamp the benefits of parallel processing. This data movement overhead is much greater for a distributed memory cluster, but these data movement effects impact shared memory computers as well.

We can very crudely model the impact of parallel overhead by adding a parallel overhead term to our equation for the run time of a parallel program. From the development of Amdahl's law, we wrote:

$$T_p = \alpha \times T_s + (1 - \alpha) \times \frac{T_s}{P}$$

The parallel overhead would appear as a new term that grows with the number of processors, $P$:

$$T_p = (\alpha + \gamma \times P) \times T_s + (1 - \alpha) \times \frac{T_s}{P}$$

where we model the overhead as a small constant $\gamma$ that is scaled by $P$. The result of parallel overhead is speedup curves such as the one in Figure 2.3. The speedup climbs quickly with a general growth constrained by the serial fraction (as predicted by Amdahl's law). As processors are added, the parallel overhead grows and eventually pulls the figure downward at a rate more extreme than would be suggested by the limitations from Amdahl's law. Sometimes you are able to create programs with little parallel overhead and an extremely small serial fraction. These cases lead to speedup curves such as the one shown in Figure 2.1. Far more often, however, your curve will look more like the one in Figure 2.3. Experienced programmers expect this shape and will conduct careful scalability studies to see where the speedup curve falls off to make sure they understand limitations to the scalability of their parallel programs.

## 2.4   Strong Scaling vs. Weak Scaling

So far in this chapter, we have considered how performance changes as the number of processors grows for a fixed-size problem. This is called *strong scaling*. Since the problem size is fixed, as the number of processors increases the amount of data for each processor to work with decreases. Eventually as the number of processors continues to grow, there may be insufficient work to keep the additional processors busy.

Strong scaling is difficult to support. This is one of the key points Amdahl was making when he published the law named after him. However, it is perhaps an overly pessimistic way to think about how we might want to use parallel computers. In many problem domains, if you pick a problem of size $N$, a specialist working in that domain can show you a problem that is even more interesting that is of size $2N$ or larger. In other words, there are usually good reasons based on the needs of a problem domain to let problem sizes grow.

Why would we want to do this? Let's return to our analysis of speedup when problems have some fraction of their execution that is serial. What if this serial

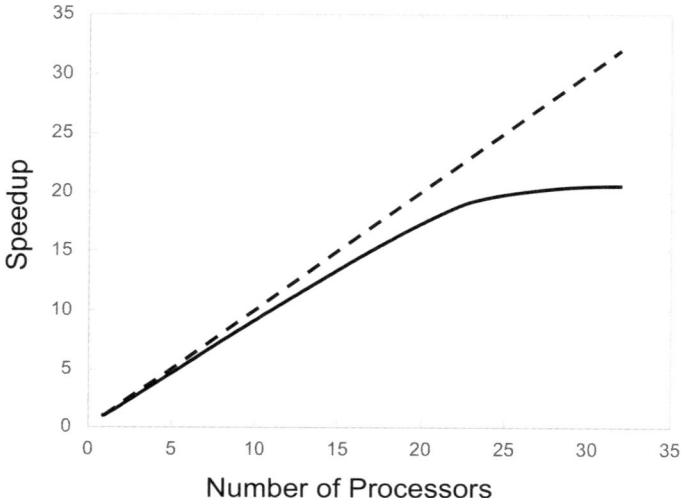

Figure 2.3: **Speedup as a function of the number of processors with a parallel overhead term** – We use linear scales for speedup *vs.* log scales for the number of processors. The dashed line is the line we would expect given perfect linear speedup. It is included to provide a visual point of reference in understanding the observed speedup data. The solid line is a speedup curve modeled as having a parallel fraction of 0.995 and a parallel overhead term of 0.0005.

fraction, $\alpha$ was not a constant. What if it were a decreasing function of the problem size, $N$? This means our speedup is now a function of both $P$ and $N$.

$$Speedup(P, N) = \frac{T_s}{T_p} = \frac{T_s}{\alpha(N) + \frac{(1-\alpha(N))}{P} T_s}$$

If the serial fraction approaches zero in the limit of N approaching some sufficiently large size ($N_{large}$) we get

$$Speedup(P, N_{large}) = P$$

It turns out for many problems, $\alpha(N)$ is indeed a decreasing function of $N$. For example, in many dense linear algebra problems, the work scales as $N^3$ where $N$ is the order of the matrices in the problem. If the serial fraction comes from setting up the matrices in the first place, that scales as the size of the matrices which is $N^2$. For sufficiently large $N$, the work term of $O(N^3)$ overwhelms any terms of $O(N^2)$ and we can ignore the contribution of the terms based on the serial fraction.

This results in scalability studies where the amount of work per processor is fixed. As processors are added, the overall problem size grows. This is called *weak scaling*. If the problem size per processor is fixed and if the parallel overhead is negligible, the time for an ideal weak scaling problem is fixed. In other words, for weak scaling, you plot run time as a function of the number of processors and the resulting curve should ideally be flat.

## 2.5   Load Balancing

We can think of the work defined by a program as the full set of operations carried out by the program. When we create a parallel version of a program, we break the work into chunks and assign those chunks to threads that execute them. If we have multiple processors so the threads can execute at the same time, we do work in parallel and reduce the overall execution time of the program.

A parallel program is done when its last thread is done. This is an important point to understand so we will repeat it, but in a slightly different way: *the slowest thread determines when all the threads are done*. Let's assume for now that all the processors run at the same speed. And let's assume for now that we are running on an SMP machine so we can ignore differences in the cost to access different regions of memory. Then a thread is "slow" when it has more work to do than the other threads.

Therefore, our goal in designing a parallel program is to have all threads finish at the same time which means we want them all to have the same amount of work to do. If you think of work as imposing a load on the threads, we say that our job as algorithm designer is to "balance the load between threads".

As we study parallel programming, we will spend a great deal of time discussing *load balancing*. We will lay out the broad view of the subject now and leave the details for later. Four distinct cases define our options for load balancing. We generate them from two pairs of opposing forces:

- *Explicit vs. Automatic*: Does the programmer work out a fixed formula to generate the balance for the load? Or does the balanced load emerge automatically in the course of the computation?

- *Static vs. Dynamic*: Is the way work is broken up and scheduled for execution fixed at compile time? Or does it happen dynamically as the program runs?

We will define the four cases for load balancing and give a brief example for each case. We do not intend, however, to belabor this topic too deeply. We want to lay the foundation for this important topic in this chapter, but the real learning will happen later when you apply these concepts in your OpenMP programs.

**Case 1, Explicit, Static**: The programmer defines the chunks based on logic in the program. These expressions are fixed at compile time and therefore as the program runs, it has a limited capacity to adjust the way the load is balanced. For example, a programmer may have four threads and break up work into four chunks selected so the work will take about the same time on each thread. Then assign one chunk to each thread and achieve an effective load.

**Case 2, Explicit, Dynamic**: The programmer writes logic into the code that determines how work is distributed, but as the program runs it pauses from time to time and revisits that logic to dynamically redistribute the load. This would happen in gravitational simulations whereas the system evolves, some regions become more populated and others less so; requiring that the way regions are mapped onto threads would need to be adjusted at run time (hence why this is considered dynamic).

**Case 3, Automatic, Dynamic**: The program creates a sequence of chunks and places them in a queue. Threads grab a chunk of work, complete it, then go back to the queue for more work. The work is dynamically balanced among the threads but at no point does the programmer have to decide which thread gets how many chunks. This approach is valuable for problems where the work is highly variable and unpredictable. It also is valuable when the processors running the threads run at different speeds (for example, in a cluster where some nodes are new and others are a generation or two behind). In this case, the faster processors just naturally pick up more work than the slower ones.

**Case 4, Automatic, Static**: The essence of a static load balancing strategy is that the distribution of the work is fixed before the work begins. In problems where the work is dependent on the input data set but the nature of the system benefits from a static work distribution, it is sometimes advantageous to have a process at run time inspect the computation to determine a static schedule that a collection of threads will then use. This is often done, for example, when solving sparse linear algebra problems on a GPU for which it is vital to understand the distribution of non-zero elements in the arrays as well as the fill-in patterns as the computations proceed.

We will encounter these cases repeatedly as we explore OpenMP, especially cases 1 (explicit, static) and 3 (automatic, dynamic).

## 2.6   Understanding Hardware with the Roofline Model

Speedup curves describe trends in a parallel program as the number of processors increases. When thinking about performance, however, you ultimately need to understand your performance relative to the raw performance available from the hardware. Instead of investigating how performance changes as you add more processors, consider the absolute speed of the computation and if it is acceptable given the algorithm and the details of the hardware.

To explore these issues, we use a roofline model [16]. The roofline model is a visual tool to help understand the limitations of a computer system relative to the features of a particular algorithm. In the model you plot performance *vs.* arithmetic intensity.

- *Arithmetic intensity* is the ratio of the number of floating-point operations (Flops) performed by a program relative to the data movement required to support those operations.

- Performance is expressed as a rate: floating point operations per second (FLOPS). When the rate of the computation is dominated by memory movement, the performance is bound by the memory bandwidth and the rate plotted becomes floating point words moved per second.

Ideally, you construct a roofline model for the specific system and algorithm you are working with. It works at a high level, however, to construct a generic roofline plot based on the peak performance of the system. We show a typical generic model in Figure 2.4.

Algorithms with a high arithmetic intensity (such as linear algebra over dense matrices and short range forces in particle methods) have so much computation to carry out for each byte accessed from memory that they are limited by the peak floating point performance of the system (Max GigaFLOPS). These computations

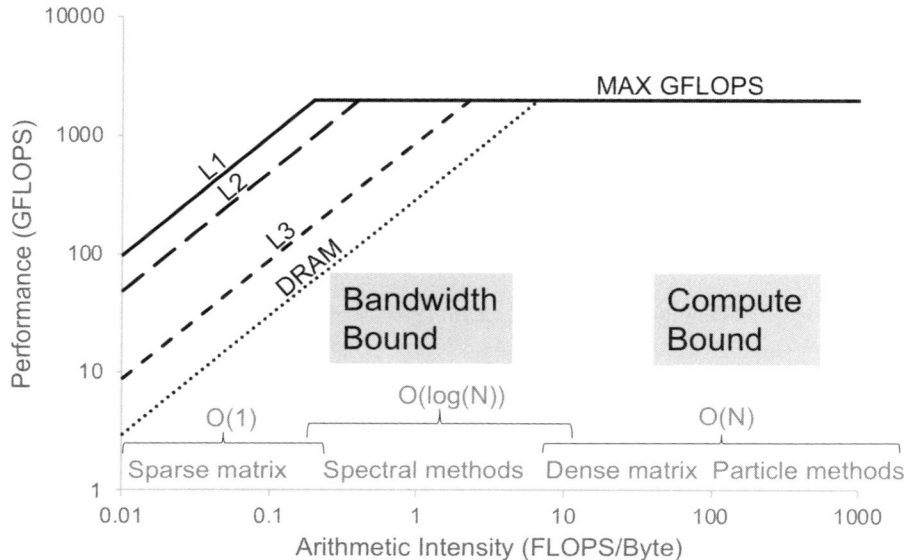

Figure 2.4: **Roofline performance model for a system with three levels of cache and DRAM memory** – The lines show the different performance limits. The upper horizontal line is the performance upper bound for programs limited by floating point operations. The slanted lines sloping downward and to the left show the performance bounds for programs limited by different levels in the memory hierarchy.

are *compute bound*. It is important to use the appropriate value for the max GigaFLOPS in the roofline plot. The value must match the precision used in the computations being studied(e.g., half, single, or double precision) and the operations that dominate the computation (e.g. scalar or vector operations or even more complex operations such as a fused multiply-add).

At the other extreme are problems of low arithmetic intensity that are bound by the movement of data across the memory hierarchy. We express the performance for the various levels of the memory hierarchy in terms of floating point numbers moved per second. The size of the memory increases but the bandwidth decreases as we go from L1 to L2 to L3 to memory (DRAM). If the problem is decomposed into blocks that fit into L1 cache and memory movement into L1 is overlapped with computation, the performance is bound by the L1 bandwidth. If L1 reuse is poor we move down to L2 or even the L3 cache; once again largely depending on how well the computation can be decomposed into blocks that fit in cache. In the worst case scenario, for a bandwidth bound computation performance is limited by the rate of data movement with main memory (DRAM).

Computations based on FFTs and spectral methods fall in the middle of the arithmetic intensity scale. With the right cache management, they can run near the peak floating point performance for the system. More typically, however, they end up being limited to some degree by the memory hierarchy.

The roofline plot is constructed based on the features of the system. Evaluate your algorithm to estimate its arithmetic intensity. Then measure the observed performance and see where you sit on the roofline plot. If you hit the peak performance available for that region of the roofline plot, you know you are done and additional optimization work is unlikely to pay off. If you fall well below the peak, however, the roofline plot suggests that it makes sense to work on different ways to restructure your computation to improve performance. The roofline plot also can be used to suggest where you might need to change your algorithms. If your arithmetic intensity puts you under the sloping, memory-bound lines, can you change your algorithm to increase the arithmetic intensity and move right into a higher performance region of the roofline plot? Your goal, in essence, is to use the roofline model to guide you on an optimization path where you move the "dot on the plot" for your performance upwards and to the right.

# 3 What is OpenMP?

OpenMP is an Application Programming Interface for writing parallel programs. While it started with a focus on multithreaded programs for SMP computers, it has grown over the years to address NUMA systems and attached devices such as GPUs. In this chapter we will explore the history of OpenMP and discuss the high level structure of this important standard.

## 3.1 OpenMP History

During the 1980s, there were a small number of shared memory computers on the market. Programmers writing applications for these machines quickly saw the need for a portable API for these systems. Several efforts were launched in the 1980s and 1990s to create such an API, but those failed.

The problem for these early standardization efforts was there just wasn't much attention paid to shared memory computers at that time. Systems based on distributed memory dominated the supercomputing scene. In these systems, there was a communication network with computers sitting at the nodes in the network. Each computer had its own memory; hence their name, *distributed memory systems*. If these distributed memory systems had shared memory nodes, these nodes only had a few processors. It was often easier to just run additional distributed memory processes on each node and ignore shared memory altogether.

In late 1995, the computing landscape for shared memory computers shifted. Intel released a chipset that supported up to four CPUs in an SMP configuration. This moved SMP computers from specialized computers for HPC to the mainstream market. It also meant that the nodes in the workstation-clusters used for high performance computing were more likely to be SMP nodes with larger numbers of CPUs. Also around that same time, SGI acquired Cray Research. Each of the two companies had their own ways to program shared memory machines. Once they became a single company, they needed to pull the two product-lines together around a shared programming model. The final and most important ingredient came from the application programmer community. They pulled together under the leadership of the U.S. Accelerated Strategic Computing Initiative (ASCI) and used their combined voice to push vendors to define a standard for programming SMP systems.

The ASCI application programmers worked closely with the major HPC vendors in late 1996 and 1997 to define OpenMP, with version 1.0 released in November

1997. Due to all the hard work from the earlier shared memory standardization efforts and drawing on the experience of the handful of vendors who had been selling shared memory systems since the 1980s, the basic design of OpenMP fell together fairly quickly. The guiding principles of our work were:

- Standardize existing practice rather than establish a research agenda. We wanted an API that vendors could implement quickly without having to go through a protracted research process.

- Support portable, efficient, and comprehensible shared-memory parallel programs.

- Provide a consistent API for Fortran, C, and C++ so programmers can easily move between languages.

- Create a small API just large enough to express important control-parallel patterns.

- Commit to strict backward compatibility so programmers will never need to rewrite code to adapt to new versions of the standard.

- Support writing code that is serially equivalent, i.e., code that produces the same result when run in parallel or when run as a serial program.

Within a year of releasing the standard, OpenMP compilers were available from all mainstream HPC, shared-memory vendors. Application programmers could write code once and just by recompiling their programs move from one shared memory computer to another. This may seem mundane by today's standards, but in the late 1990s when OpenMP became available, it was revolutionary.

We knew when we created OpenMP that it needed to be a "living" language that could evolve as hardware and algorithms change. Therefore, we created a non-profit corporation to own the standard, protect it from inappropriate manipulation by any single vendor, and guide its continuous development. This organization is called the *OpenMP Architecture Review Board* or the ARB. Under the leadership of the ARB, we have continued to produce new versions of the specification. We summarize the development in Figure 3.1 where we show the page count for the specification (i.e., ignoring front material and the appendices).

OpenMP started with version 1.0 for Fortran in 1997. It targeted basic loop-level parallelism and required only 40 pages to fully define. We continued refining and

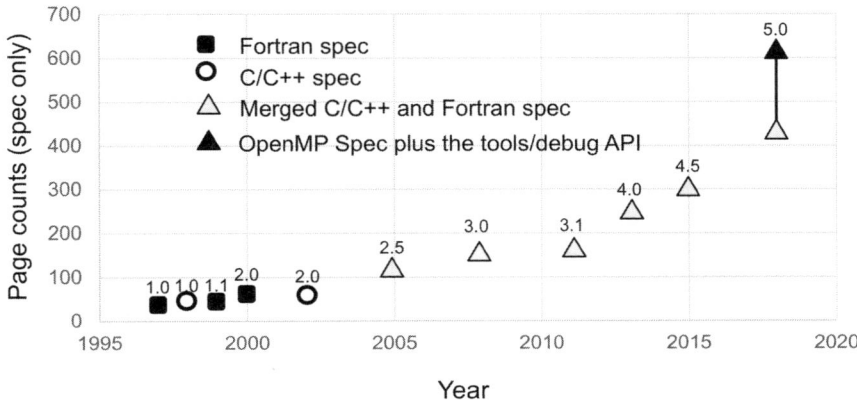

Figure 3.1: **Page counts for OpenMP Specifications over time** – These page counts exclude front-matter and appendices. They only cover the parts of the specification that define OpenMP. For OpenMP 5.0 we distinguish between page count for the API alone and page count that includes the new tools interface.

developing new features on two tracks: one for Fortran and another for C/C++. These two tracks were "serial" since the same people were working on the Fortran and C/C++ Specifications. So if we found a problem in OpenMP while working on, for example, the Fortran specification, it could be 3 or 4 years before the required changes showed up in C/C++ since we had to finish the Fortran specification before we could move any corrections into the C/C++ document. To fix this problem, we merged the C/C++ and Fortran Specifications into a single document. This was OpenMP 2.5 released in 2005 with 117 pages.

From OpenMP 1.0 to 2.5, the focus remained on parallelism expressed in terms of teams of threads working together on parallel loops. For OpenMP 3.0 released in 2008 with 151 pages, we added tasks to OpenMP. With tasks, we could move beyond regular loops and consider recursive and other irregular algorithms.

In OpenMP 4.0 (released 2013 with 248 pages) we added a host device model. Using this model, programmers could express algorithms for data parallel, throughput-optimized processors such as GPUs. This was a huge break from the prior art in OpenMP and moved us into the complex world of systems with multiple address spaces. We also added in OpenMP 4.0 ways to explicitly control execution on the vector units of a processor, so-called SIMD or Single Instruction Multiple Data parallelism (which we discuss in detail in Section 12.2).

In November of 2018, the ARB released OpenMP 5.0. It is huge with over 600 pages of which 437 defines the API and the rest specifies a new tools interface. It expands the functionality of tasking in OpenMP and updates the API for the latest developments in C++ and Fortran. It adds additional enhancements across the API with better support for managing complex memory hierarchies, expanded GPU support, iterators to support modern C++ styles of programming, and much more.

This discussion has touched on the main features in the evolution of OpenMP. Our goal was not to provide a detailed history (which can be found in the paper from [2]); rather, we wanted to show how OpenMP, as a living language growing in complexity over 20+ years, has become quite complex. When we started, our goal was to make parallel programming as simple as we could, which is still the case since OpemMP is backward compatible. With the large size of the 5.0 specification, however, finding the simple core of OpenMP that we started with has become increasingly difficult.

## 3.2    The Common Core

We believe that we need to change how we teach OpenMP. Rather than following the progression of the specification, which is how we used to teach OpenMP, we should isolate the core of OpenMP that captures the inherent simplicity of the API. It turns out that most programmers rarely if ever move beyond that core of OpenMP: those essential elements used by the majority of OpenMP programs. We call this the *OpenMP Common Core*. You can find the items from OpenMP that are part of the Common Core in Table 3.1.

We have not discussed these concepts yet so do not spend any time trying to understand them. That will come later as you go through the rest of the book. We list them here to provide a roadmap for our journey through OpenMP's Common Core.

## 3.3    Major Components of OpenMP

Before we complete our overview of OpenMP and start our detailed exploration of the OpenMP Common Core, we want to close with a description of the high level structure of OpenMP and how it fits in with a typical shared memory computer. We show this in Figure 3.2.

Table 3.1: **The pragmas, runtime library functions, and clauses that make up the OpenMP Common Core plus the associated fundamental concepts of multithreaded computing.**

| OpenMP pragma, function, or clause | Concepts |
| --- | --- |
| #pragma omp parallel | Parallel regions, teams of threads, structured blocks, interleaved execution across threads |
| int omp_get_thread_num() <br> int omp_get_num_threads() <br> void omp_set_num_threads() | The SPMD Pattern: Create a parallel region and split up the work using the number of threads and thread IDs |
| double omp_get_wtime() | Timing blocks of code, Speedup, and Amdahl's law |
| export OMP_NUM_THREADS=N | Internal control variables and setting the default number of threads with an environment variable |
| #pragma omp barrier <br> #pragma omp critical | Implications of interleaved execution, race conditions and synchronization |
| #pragma omp for <br> #pragma omp parallel for | Worksharing, parallel loops, loop carried dependencies |
| reduction(op: list ) | Reductions of values across a team of threads |
| schedule( static   [, chunk]) <br> schedule(dynamic) [,chunk]) | Loop schedules, loop overheads and load balance |
| private( list ) <br> firstprivate ( list ) <br> shared( list ) | The OpenMP Data environment: the default rules and the clauses that modify default behavior |
| default (none) | Force explicit definition of each variable's storage attribute |
| nowait | Disabling implied barriers on worksharing constructs, the high cost of barriers, and flushing memory |
| #pragma omp single | Work done by a single thread |
| #pragma omp task <br> #pragma omp taskwait | Tasks, Task completion and the data environment for tasks |

We begin at the bottom with the hardware. The OpenMP Common Core assumes an SMP model of the shared memory computer. It does not include any of the more advanced OpenMP features added to address NUMA computers. Above the hardware sits the *System Layer*. The Operating system supports the shared memory computer with some sort of threading model. OpenMP uses whichever threading model the operating system provides, which in most cases is pthreads. Above the OS layer is the *OpenMP runtime system*. This consists of the low level libraries and software components that support the execution of OpenMP programs. It is not defined by the OpenMP specification. It is written by the implementor of OpenMP.

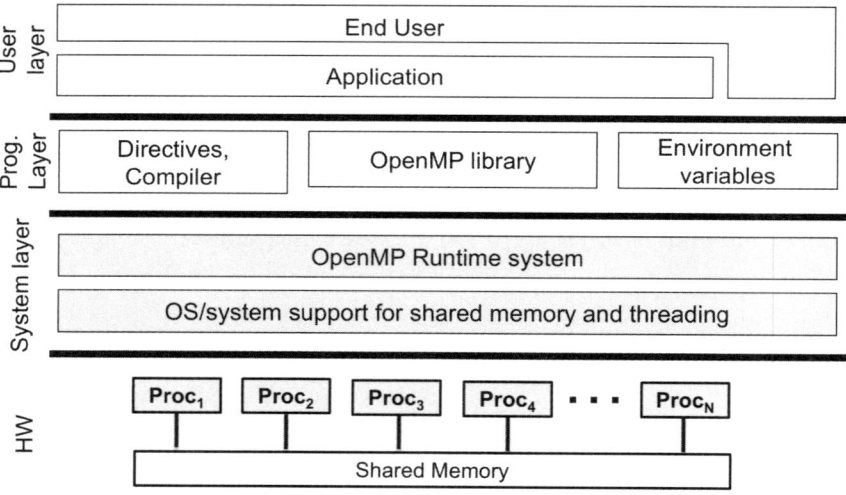

Figure 3.2: **The OpenMP "solution stack" for the Common Core.**

The next layer is the *Programming layer*. This is the OpenMP API defined by the OpenMP specification. It consists of three basic groups of items.

- **Directives and an OpenMP aware compiler**: the directives tell the compiler what to do on behalf of the OpenMP programmer to create a multithreaded program.

- **OpenMP library**: functions defined by the API for interacting with the computer as a program executes. It addresses issues that cannot be resolved at compile time such as the number of threads as well as low level primitives to control execution of a program.

- **Environment variables**: control features of an executing program and set default parameters at runtime.

The final layer is the *user layer* where by "user" we mean the person running (rather than creating) an OpenMP program. The shape of the "End User" boxes at the top layer shows that while the application program interacts with the full range of OpenMP items, a user may directly interact with the OpenMP runtime through the environment variables.

This closes our overview of OpenMP and its history. You now have all the context you need to start your exploration of the OpenMP Common Core.

# II THE OPENMP COMMON CORE

In this part of the book, we present the OpenMP Common Core. It includes the directives, clauses, library routines, environment variables, and associated concepts that the overwhelming majority of OpenMP programmers use all the time.

We begin in Chapter 4 with an overview of OpenMP and the basic mechanisms to manage threads in an OpenMP program. A surprisingly broad range of parallel algorithms can be covered with these basic thread management features from OpenMP.

Then we move to Chapter 5 to cover what most programmers consider the "bread and butter" of OpenMP: parallel loops. Using worksharing-loop constructs, we find the compute intensive loops in our code and split the loop iterations across a team of threads. It suggests a straightforward, incremental approach to parallelism as we move from one loop to the next; parallelizing code until our performance goals are reached. Incremental, loop parallelism in many cases does not lead to optimal performance. To get the most from a parallel algorithm, you often need to fuse loops, restructure data to optimize memory movement, or completely change the underlying algorithms. Incremental loop parallelism is a great way to get started with OpenMP, however, and in many cases, it is all programmers are willing to do.

Once you have threads running within a single address space, you quickly run into issues arising from how data is shared between threads. This is the topic of Chapter 6 where we discuss the OpenMP data environment. This topic is extremely important since many of the bugs in OpenMP programs arise from mistakes in managing the data environment.

The next piece of the Common Core is task-level parallelism: the topic of Chapter 7. Parallel loops and explicitly managed threads cover a wide range of algorithms. They can be challenging, however, for cases where the parallelism is not found in `for-loops` but in less regular structures such as `while-loops` with an unknown a-priori length or recursive algorithms. In these cases, the parallelism is best addressed through dynamic task queues managed by a team of threads. This style of parallelism has grown rapidly as applications of parallel computing move beyond the basic differential equation solvers at the core of scientific computing.

In Chapter 8, we explore the other leading source of challenging bugs in multi-threaded programs: the rules that govern which values can be returned when a variable shared between threads is read. This set of rules is called a *memory consistency model*. We explain the simplified memory model used in the OpenMP common core and the limitations of this model.

We close this part of the book with a quick recap of the Common Core in Chapter 9. Based on many years of experience with OpenMP, we believe most programmers rarely need items from OpenMP that are missing from the Common Core. Therefore, master the Common Core and then, only if needed, go beyond the Common Core.

# 4 Threads and the OpenMP Programming Model

## 4.1 Overview of OpenMP

We created OpenMP with a particular use case in mind. The starting point is a sequential program written in C, C++, or Fortran. The programmer's goal is to convert this sequential program into a parallel program to run on a shared memory multiprocessor computer. Ideally, this would be done with minimum disruption to the original code.

A minimally intrusive way to modify code is through compiler directives. A compiler directive tells the compiler to *do something* on behalf of the programmer. In most cases, the directives in OpenMP are semantically neutral and do not change the meaning of a program. We made this choice in the design of OpenMP so programmers could write code that was semantically equivalent when built with an OpenMP-enabled compiler to be run in parallel, or when built with a compiler unaware of OpenMP to be run as a serial program.

As you will see, we were successful and this semantic equivalence between serial and multithreaded execution is possible. It is not, however, required. You can write code that only runs correctly when executed with multiple threads. OpenMP does not protect you from your choices as a programmer. We urge programmers, however, to adopt a programming discipline where the parallel code is semantically equivalent to the original serial code.

## 4.2 The Structure of OpenMP Programs

The directives defined by OpenMP take different forms depending on the host programming language. We show the form for the directives in C, C++, and Fortran in Table 4.1. In C and C++ the directive is expressed as a pragma. In Fortran, we use a special form of a comment statement to indicate a compiler directive. The names used in the directives are almost always the same for Fortran, C, and C++. This lets a programmer easily move between languages.

Most of the OpenMP directives apply to a block of code. The compiler, based on the directive, does something to the block of code associated with a directive. This usually involves "outlining the code" which is "compiler jargon" for when a compiler creates a function from statements in a program during compilation. The programmer never sees this function explicitly. The compiler creates the function and calls that function in the code it generates during the compilation process.

Since the code in the block associated with a directive will be turned into a function, the compiler must be able to make some assumptions about the code. In particular, the compiler needs to assume that barring an error condition that shuts down the program (e.g., a call to `exit` from C/C++ or a `STOP` statement in Fortran), the block of code will be entered from the top of the block and exit from the end of the block. In other words, the program does not jump into the middle of the block or jump out of the middle of the block. OpenMP calls this block of code a *structured block*.

Table 4.1: **General form of directives in C/C++ and Fortran** – The combination of a directive and a structured block is called a *construct*.

| **C/C++ directive format and an example with a structured block** |
|---|
| **#pragma omp parallel** *[clause[[,] clause]...]* |
| **#pragma omp parallel** `private(x)`<br>`{`<br>`  ...   code executed by each thread`<br>`}` |
| **Fortran directive format and an example with a structured block** |
| **!$omp parallel** *[clause[[,] clause]...]* |
| **!$omp parallel** `private(x)`<br>`  ...   code executed by each thread`<br>**!$omp end parallel** |

The basic syntax of directives in OpenMP is shown in Table 4.1. C and C++ are block structured languages. The language definition includes the concept of a block where a block is one line of code or multiple lines of code between curly braces { and }. Hence, we can use the features of C and C++ to define the structured block associated with an OpenMP construct. Fortran, however, is not block structured. For Fortran, we need to add a directive to indicate the end of a block. As with C and C++, a block in Fortran is one statement or a set of statements between the opening directive and a directive that terminates the block (e.g., the `!$omp end parallel` in Table 4.1).

While you use a curly brace to close a block in C/C++ as opposed to Fortran's end directive, it can be confusing when many blocks are nested (a practice that you will see is quite common). A code sequence can present a programmer with numerous closing curly braces and it can be confusing to match the closing curly brace to the right construct. Hence, we often add a comment to a curly brace to specify which OpenMP pragma it binds to.

```
#pragma omp parallel
{
      .... do lots of stuff
} // end of parallel region
```

This is not required and we often skip this rule in our own code. In complicated programs with many blocks, however, this simple comment can help you understand your code.

We use two terms in OpenMP to describe the code executed by an OpenMP program. A directive and its structured block is called a *construct*. Some OpenMP literatures use the term "lexical extent" to refer to the code visible in the compilation unit containing the directive. A construct is static; it's what you see in the source code containing a directive. We use the term *region* to refer to all the code executed by a thread associated with a construct. It includes the code defined in the compilation unit with the directive but also the code inside any functions called from within the structured block. The region is the "dynamic extent" of a construct since you do not necessarily know which code is executed inside a construct until the program runs.

A compiler directive defines transformations made by the OpenMP compiler during compilation of a program. There are features of a parallel computation that can only be assessed at runtime. For example, a directive may indicate (as we will soon discuss) that a number of threads should be created. The code to create those threads is known at compile time, but the system does not know how many threads will actually be used when a program runs. This number depends on the hardware, the number of cores associated with the system, and the number a user might request when running the program.

Features of a program only known at runtime are referenced inside an OpenMP program through a set of library routines. For example, we will shortly learn about a runtime library routine that returns the number of threads in use at a point in the program (`int omp_get_num_threads()`) and another that indicates the thread rank or ID (`int omp_get_thread_num()`).

Finally, there are aspects of a computation that arise when an executable runs on a system. For example, the user of a program may want to change the default number of threads or describe preferred ways for loops to be divided among a set of threads. These execution-time issues are handled through a set of environment variables.

That completes our look at the basic structure of OpenMP. It does not take much to get started with OpenMP since our goal in creating OpenMP was to provide a simple API for application programmers. To learn OpenMP, you just need to master a modest number of elements from the OpenMP specification:

- compiler directives

- runtime library routines

- environment variables

In addition, you need to remember how the code in your program interacts with the elements of OpenMP, in particular, the idea of a construct (a directive plus the code directly visible in the compilation unit containing the directive ) and a region (code in the construct plus any code in functions called from within a construct's structured block).

## 4.3   Threads and the Fork Join Pattern

As we saw in Section 1.2 (with the *Hello World* program), in OpenMP, you create threads with the `parallel` construct defined in Table 4.2.

Table 4.2: **The parallel construct in C/C++ and Fortran** – This construct will create (or fork) a team of threads, execute the code inside the construct, and then join the threads together so only the master thread continues.

| C/C++ | Fortran |
|---|---|
| **#pragma omp parallel** | **!$omp parallel** |
| { | |
| `...code executed by each thread` | `...code executed by each thread` |
| } | **!$omp end parallel** |

The `parallel` directive causes a collection of threads to be created. We call these threads a *team of threads*. At the end of the construct the team of threads is destroyed and only one thread, the thread that first encountered the `parallel` directive, continues execution.

The fundamental pattern implied by this construct is called the *fork-join pattern*. As shown in Figure 4.1, the program starts with an *initial thread*. When this thread encounters code that could benefit from parallel execution, a parallel construct "forks" a team of threads to execute the code of this *parallel region*. Notice that the initial thread is part of the team. We call this the "master thread" of the team. The team of threads does the work defined by the parallel region. When done, the team of threads "join" together, all threads other than the master are "destroyed" and the master thread continues and executes a *sequential part* of the program. Later the master thread may see yet another block of code that can benefit from parallel execution. It will then create a new team with perhaps a different number of threads, and continues with a new parallel region until the team finishes its work and joins. We can even nest teams of threads.

There are a couple of low level details to clear up. When we say "destroyed", we use quotes for a reason. A good implementation of OpenMP may choose to reduce the overhead of creating and destroying threads by moving them back and forth between a lower level structure called a thread pool. It appears to the programmer that all the threads other than the master thread are destroyed, but behind the scenes they may not be destroyed at all. This usually does not matter and you can ignore this low level implementation detail. There are a few corner cases, however, where this manner of handling threads after the join may have significant performance implications.

The second detail we need to clear up concerns the timing of events around the join. All of the threads must execute the join before the one remaining thread continues. You can safely assume that once a thread continues past a join, the threads in the team have completed their work and shipped any results back to memory. This behavior at a join appears often in the OpenMP program. We call this a *barrier*. A *barrier* is a point in a program that all threads in a team must reach before any threads continue. In this case, all threads reach the barrier, then they execute the join, and the initial thread continues. We will have much more to say about barriers when we discuss synchronization later in the book.

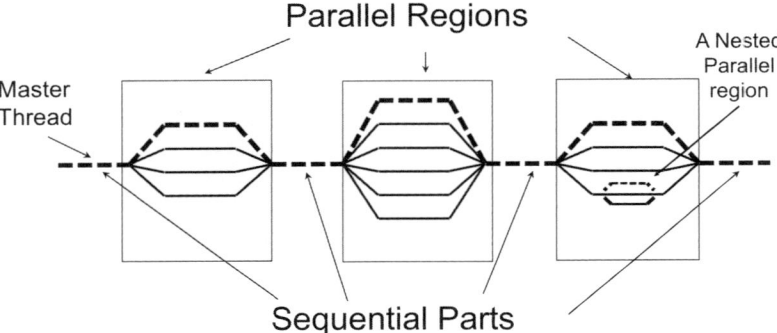

Figure 4.1: **The fork-join model** – A program starts as a single thread. It creates (or *forks*) a team of threads each of which executes a block of code. When done, the threads *join* (i.e., are destroyed) and the single original thread continues.

The OpenMP Common Core is an explicit API[1]. The directives do not make *suggestions* to the compiler. They *tell* the compiler what to do as it generates a parallel program. Also, the threads are explicitly exposed. There is a team of threads and the programmer can know how many threads are in the team and can distinguish between threads.

A program starts with a specific value for the default number of threads to use when creating a team of threads. An implementation of OpenMP, however, has a great deal of flexibility in choosing that number. The most common choice is to set the default number of threads in a team to the number of cores seen by the operating system. For large multiprocessor systems with hundreds or thousands of cores, the system may set a lower default. Or for a system shared between many users, the default may be set lower than the number of cores to prevent people from being greedy and hogging the whole machine for their programs. The key point is that an implementation of OpenMP has the flexibility to set a default value for the number of threads in a team based on what makes the most sense.

Of course, OpenMP provides a way for programmers to modify the default number of threads to request when a team of threads is created. There are several ways to

---

[1]OpenMP with version 5.0 has added constructs that are declarative. These declarative constructs describe what they want the system to accomplish, but say nothing about "how" the work will be carried out. As OpenMP grows and works to address the needs of heterogeneous systems, it is likely that additional declarative directives will be added.

do this, but for now we will focus on just one: using a library routine to explicitly change the default value. We show this in Table 4.3.

Table 4.3: **Library routine to change the default number of threads in C/C++ and Fortran** – this default number is used as the number of threads requested for subsequent parallel regions until explicitly overridden by a new value.

| |
|---|
| void **omp_set_num_threads**(int num_threads); |
| **subroutine omp_set_num_threads**(num_threads)<br>integer num_threads |

Consider the code in Figure 4.2. It contains a single parallel region. A team of threads is forked. It will consist of up to 4 threads. Even though the programmer requested 4 threads, the system may choose to create a team with fewer than 4 threads. You can depend on one thing, though: the size of a team of threads, once formed, is fixed. The number of threads a team starts with is the number of threads it will have when it completes its work and executes the join at the end of the parallel region. An OpenMP runtime system will not reduce the size of a team of threads once launched.

To do anything useful with threads, you need to understand how the data environment interacts with the threads. The rules can get complicated. Therefore, we will later dedicate an entire chapter to this topic. We can do a great deal of OpenMP programming, however, with the simple default rules for the OpenMP data environment.

The basic rule is that variables declared prior to the parallel construct are shared between threads while variables declared inside the parallel construct are private to a thread. Consider the code in Figure 4.2. We have a simple function called `pooh()` that assigns an input integer to a location in an array. Looking at the main function, an array is statically declared; that is, we provide a fixed dimension for the array when it is declared. In this case, we initialize the values in this array to zero. Then after requesting a particular number of threads to use in subsequent parallel regions (i.e., we set the default number of threads to request when creating teams) we enter the parallel region. Since the array is declared by the initial thread prior to the parallel region, it is visible to all the threads inside the parallel region. All threads will see the same array, `A`.

Inside the parallel region, we declare an `int` variable `ID`. C is a block structured language with well-defined scopes for variables. If a variable is declared inside a

```
1   #include <stdio.h>
2   #include <omp.h>
3
4   // a simple function called by each thread
5   void pooh(int ID, double* A)
6   {
7      A[ID] = ID;
8   }
9
10  int main()
11  {
12     double A[10] = {0};      // an array visible to all threads
13     omp_set_num_threads(4);
14     #pragma omp parallel
15     {
16        int ID = omp_get_thread_num();  // a variable local to each thread
17        pooh(ID, A);
18        printf(" A of ID(\%d) = \%lf\n",ID,A[ID]);
19     } // end of parallel region
20  }
```

Figure 4.2: **Data movement and parallel regions** – This simple program sets the default number of threads to request for a parallel region to 4. A parallel region is defined within which a thread ID is set and a simple function is called. Key points form this program: (1) all the threads independently execute the same block of code in this parallel region, (2) all threads have access to the array declared prior to the parallel region, and (3) each thread has its own, private copy of the `int ID`.

block, then it is local or *private* to the block. Hence, every thread will get its own variable named `ID`. We then call a runtime library routine to provide a number to uniquely define a thread (as defined in Table 4.4). The thread number is a rank, that is, a number that ranges from zero (for the master thread) to the number of threads minus 1.

Table 4.4: **Library routine to return the thread number in C/C++ and Fortran** – The returned value is the thread rank within a team and ranges from zero to the number of threads minus 1.

| int **omp_get_thread_num**(); |
|---|
| integer **function omp_get_thread_num**() |

Finally, there are times that you want to know how many threads you have in a team. Since you cannot assume the number you requested is the number you have, we need a way to query the system to get this number. We show how to do this in Table 4.5.

Table 4.5: **Library routine to return the number of threads in the current team in C/C++ and Fortran**

| |
|---|
| int **omp_get_num_threads();** |
| integer **function omp_get_num_threads()** |

These three library routines used to manage threads are critically important in OpenMP. Just to make sure you understand them in detail, we present one more example of how to use them in Figure 4.3. Four threads are requested for any parallel region following the call to omp_set_num_threads(). This is the default number of threads to use in parallel regions and it holds until a different number is explicitly set. Inside the parallel region, we query the OpenMP runtime to find the number of threads in the current team (i.e., the number of threads we *actually* got) and the thread ID.

```
1   #include <stdio.h>
2   #include <omp.h>
3
4   int main()
5   {
6     omp_set_num_threads(4);
7     int size_of_team;
8     #pragma omp parallel
9     {
10        int ID = omp_get_thread_num();
11        int NThrds = omp_get_num_threads();
12        if (ID == 0) size_of_team = NThrds;
13     } // end of parallel region
14     printf("We just did the join on a team of size \%d", size_of_team);
15  }
```

Figure 4.3: **Library routines to manage threads** – This program shows how to set the default number of threads to request in parallel regions, query the number of threads in a team, and set a unique thread ID. Notice the care taken to avoid a data race when assigning to size_of_team.

An important subtle point is exposed in Figure 4.3. What if you want to know how many threads were used inside a parallel region *after* the parallel region has completed its join and ceased to exist? C is a block structured language. After the block has finished, any variables declared inside the block (such as NThrds) go out of scope and are no longer available. The solution is to declare a variable prior to the parallel region to make it a shared variable, in this case size_of_team, and then inside the parallel region, assign it to the NThrds (the number of threads).

Notice that in Figure 4.3, only one thread is allowed to do that assignment. Since all threads see the same size for the team, couldn't we have just let every thread assign their value to the shared variable size_of_team?

Multiple threads that interleave write statements to a shared variable can collide with each other. As one thread is writing its value to the shared variable, a different thread in the team may try to execute a write to that same variable. This is called a *data race*. While for some processors (such as the x86 CPUs from Intel) the low level rules for how writes are executed guarantees that interleaved write statements will execute correctly, other processors may not provide that guarantee. Since interleaved operations can occur in any order depending on how the threads are scheduled, the final value that resides in size_of_team is undefined.

Data races lead to undefined values and depending on the program, they can make it impossible to unambiguously define the behavior of the program. Therefore, OpenMP and other programming languages (including C and C++) stipulate that any program that contains a data race is undefined. The compiler is not required to create a well-defined program in the face of race conditions.

Therefore, a key part of working with threads coded with any language (not just OpenMP) is to make sure there are no data races. Any time writes occur to a shared variable, the programmer must assure that those writes do not overlap with writes to that same variable from other threads. We will say a great deal more about data races and data movement when we discuss synchronization and the OpenMP memory model.

## 4.4   Working with Threads

We have presented a single OpenMP construct (parallel) and a trio of OpenMP routines to help manage the threads in a team. That is only a tiny fraction of what is available from OpenMP. It is enough, however, to do some serious parallel programming. In this section of the book, we will explore parallel programming by looking at different solutions to the problem in Figure 4.4. We will work with a program that estimates a value of a definite integral by approximating the area under a curve as a sum of rectangles. The integrand and range of the integration are selected so the result from this integral should be equal to Pi.

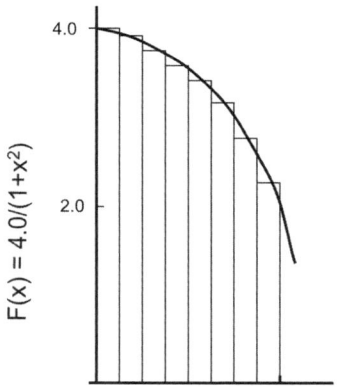

Mathematically, we know that:

$$\int_{0}^{1} \frac{4.0}{(1+x^2)}\, dx = \pi$$

We can approximate the integral as a sum of rectangles:

$$\sum_{i=0}^{N} F(x_i)\Delta x \approx \pi$$

Where each rectangle has width $\Delta x$ and height $F(x_i)$ at the middle of interval i.

Figure 4.4: **Numerical integration** – An integral can be approximated by filling in the area under a curve with rectangles and summing their areas. We choose the integrand and the limits of integration so the result should approximate Pi.

A C program that implements the algorithm defined in Figure 4.4 is presented in Figure 4.5. To keep things as simple as possible we fix the number of steps (`num_steps = 100000000`) as a file scope variable. We compute a step size and then run over the steps of the numerical integration in a `for` loop. For each iteration, we compute the center of that particular step (x in line 16) and then accumulate the value of the integrand at that position (into `sum` in line 17). After the loop has completed we multiply by the step size and if all went well, the result should approximately equal Pi.

We did add one more OpenMP concept in Figure 4.5. The goal of parallel programming is to make a program run in less time. Therefore, as we work with different ways to parallelize our Pi program, we need to track how long it takes. We do this with the OpenMP wallclock timer defined in Table 4.6. You call the timer

```
1   #include <stdio.h>
2   #include <omp.h>
3   static long num_steps = 100000000;
4   double step;
5   int main ()
6   {
7       int i;
8       double x, pi, sum = 0.0;
9       double start_time, run_time;
10
11      step = 1.0 / (double) num_steps;
12
13      start_time = omp_get_wtime();
14
15      for (i = 0; i < num_steps; i++){
16          x = (i + 0.5) * step;
17          sum += 4.0 / (1.0 + x * x);
18      }
19
20      pi = step * sum;
21      run_time = omp_get_wtime() - start_time;
22      printf("pi = \%lf, \%ld steps \%lf, \%lf secs\n ",
23                  pi, num_steps, run_time);
24  }
```

Figure 4.5: **Serial program to numerically estimate a definite integral using the midpoint rule** – The loop iterations are independent other than the summation into sum.

before (line 13) and after (line 21) the section of code being timed. Their difference is the elapsed time to execute the code in question.

Table 4.6: **Library routine to return the time in seconds from some point in the past in C/C++ and Fortran**

| double **omp_get_wtime()**; |
| --- |
| double precision **function omp_get_wtime()** |

### 4.4.1   The SPMD Design Pattern

To parallelize the program in Figure 4.5, we fork a team of threads and then have each thread handle a subset of the loop iterations. Each thread will execute the same code defined by the structured block associated with the parallel construct. Hence, we need a way to encode in the program logic that different threads should handle different loop iterations. We do this with the Single Program Multiple Data (SPMD) design pattern:

- Launch two or more threads that execute the same code.

- Each thread determines its ID and the number of threads in the team.

- Use the ID and the number of threads in the team to split up the work between threads.

As discussed in [9], the SPMD pattern is heavily used in message passing programs written with MPI and works well with OpenMP as well. It is probably the most commonly used design pattern used in parallel computing.

In the serial Pi program in Figure 4.5 we see that there is a single loop. We can split up the loop between the threads using a trick well known to MPI programmers. We call this a cyclic distribution of loop iterations. Given a rank for the thread ID (that is, integer thread identifiers that range from 0 to the number of threads minus 1), we start the loop iteration at the thread ID and increment the loop by the number of threads:

```
ID = omp_get_thread_num();
numthreads = omp_get_num_threads();
for (int i = ID; i < num_steps; i = i + numthreads) {
    // body of the loop
}
```

If there are 4 threads, for example, thread number 0 will execute iterations $0, 4, 8, 12...$, thread number 1 will execute the iterations $1, 5, 9, 13, ...$ and so forth for the remaining threads. As you can see, between the full set of threads, all of the loop iterations will be covered. However, the program still won't work since there are a few data races. We will address the data races in two steps. First, let's take care of variables that need to be private to each thread. Looking at the code in Figure 4.5, we see that the variables i and x need to be local to a thread. We can make them local by declaring them inside the parallel construct before the loop. Likewise, the variables used to manage the threads (ID and numthreads) must be private to each thread.

The interesting challenge is `sum`. Each thread needs its own copy of the sum to accumulate the sum over rectangles for the assigned iterations. We could create a private copy of `sum`, but then when the parallel region is done and the structured block is complete, the variable goes out of scope and is no longer available. We need access to those values, however, so we can combine the sums from each thread to get the final full summation. How do we create a shared variable with room to give each thread a private region of memory for the local sums? An array gives us the behavior we need. If we promote the scalar `sum` in the serial program into an array with an array element for each thread, we get the behavior we need.

Putting all these elements together results in the parallel Pi program in Figure 4.6. The cyclic distribution of loop iterations between threads is a particularly simple way to divide up a loop between threads.

There is a second approach to distributing loops across threads based on a block decomposition of the loop. We provide just the parallel construct using this approach in Figure 4.7 for our numerical integration, Pi program. The idea is to give each thread an approximately equal sized block of loop iterations. This block size is found by dividing the number of steps in the integration by the number of threads. We then multiply that block size by the thread `id` to find the beginning of the block for each thread and by `id+1` to find the index for the last iteration of the block. We need to account for cases where the number of steps in the integration is not evenly divisible by the number of threads. We handle those cases by setting the loop iteration for the end of the block for the last thread (`id = numthreads - 1`) to the number of steps (line 13).

The cyclic and blocked distributions are just one of many ways to divide the iterations of a loop among a set of threads. We will discuss the performance implications, in the cases where there are any, in the next chapter when we discuss loop-level parallelism.

For the program based on the cyclic loop distribution (in Figure 4.6), we will consider the performance of the program as we change the number of threads. We run the program on a dual-core laptop[2] with hyperthreading enabled. This means each core has two hardware threads and the operating system considers this to be a laptop with four processors. The performance results are shown in Table 4.7. The performance improves considerably moving from one thread to two, but it slows

---

[2] Intel® compiler (icc) with default optimization level (O2) on Apple OS X 10.7.3 with a dual core (four HW threads) Intel® Core™ i5 processor at 1.7 GHz and 4 GByte DDR3 memory at 1.333 GHz

```
 1   #include <stdio.h>
 2   #include <omp.h>
 3
 4   #define NTHREADS 4
 5
 6   static long num_steps = 100000000;
 7   double step;
 8   int main()
 9   {
10       int i, j, actual_nthreads;
11       double pi, start_time, run_time;
12       double sum[NTHREADS] = {0.0};
13
14       step = 1.0 / (double) num_steps;
15
16       omp_set_num_threads(NTHREADS);
17
18       start_time = omp_get_wtime();
19       #pragma omp parallel
20       {
21           int i;
22           int id = omp_get_thread_num();
23           int numthreads = omp_get_num_threads();
24           double x;
25
26           if (id == 0) actual_nthreads = numthreads;
27
28           for (i = id; i < num_steps; i += numthreads){
29                   x = (i + 0.5) * step;
30                   sum[id] += 4.0 / (1.0 + x * x);
31           }
32       } // end of parallel region
33       pi = 0.0;
34       for (i = 0; i < actual_nthreads; i++)
35           pi += sum[i];
36
37       pi = step * pi;
38       run_time = omp_get_wtime() - start_time;
39       printf("\n pi is \%f in \%f seconds \%d thrds \n",
40                           pi, run_time, actual_nthreads);
41   }
```

Figure 4.6: **SPMD parallel numerical integration with cyclic distribution of the loop iterations** – The program computes the area of the curve defined by the integrand by filling the area under a curve with rectangles and summing up their areas. This version of the program promotes the accumulation variable sum to an array and uses a cyclic distribution of loop iterations between threads.

```
1   step = 1.0 / (double) num_steps;
2   #pragma omp parallel
3   {
4       int i;
5       int id = omp_get_thread_num();
6       int numthreads = omp_get_num_threads();
7       double x;
8
9       if (id == 0) actual_nthreads = numthreads;
10
11      int istart = id     * num_steps/numthreads;
12      int iend  = (id+1) * num_steps/numthreads;
13      if (id == (numthreads-1)) iend = num_steps;
14
15      for (i = istart; i < iend; i++) {
16          x = (i + 0.5) * step;
17          sum[id] += 4.0 / (1.0 + x * x);
18      }
19  } // end of parallel region
```

Figure 4.7: **SPMD parallel numerical integration with block decomposition of the loop iterations** – The parallel region from the code in Figure 4.6, but replacing the cyclic distribution of loop iterations with a block distribution.

down with the third thread and shows very little improvement going to the fourth thread.

Table 4.7: **Run times and speedups for the numerical integration program in Figure 4.6** – Reported run times are elapsed wallclock times in seconds for the SPMD algorithm. The speedup is relative to the serial program run time of 1.83 seconds.

| Threads | SPMD | Speedup |
|---------|------|---------|
| 1       | 1.86 | 0.98    |
| 2       | 1.03 | 1.78    |
| 3       | 1.08 | 1.69    |
| 4       | 0.97 | 1.89    |

Collecting data such as that found in Table 4.7 is important when writing multithreaded programs. This means running a program multiple times and increasing the number of threads each time until you run out of cores or the speedup levels off. We have only discussed one way to change the number of threads requested for a program: using the `omp_set_num_threads()` function. This requires that you edit the source code and recompile the program anytime the number of threads changes. This can become cumbersome during a performance study.

OpenMP defines a second way to change the default number of threads. You can set an environment variable `OMP_NUM_THREADS` to change the default number of threads to be used with a program[3]. The following command issued at the Bash-shell prompt on a Linux-based system[4] would set the default number of threads in OpenMP programs to 4:

```
export OMP_NUM_THREADS=4
```

An OpenMP program checks this environment variable as it starts up and, until overridden inside the program (by a call to `omp_set_num_threads()` for example) uses that value for the number of threads to request each time a parallel construct is encountered. This makes it easy to run a program with many different numbers of threads without recompilation for each case.

### 4.4.2   False Sharing

The performance of our basic SPMD Pi program in Figure 4.6 is not very impressive. While we did not compute a serial fraction, we expect it to be quite small since by far the most time-consuming part of the computation has been parallelized. With four hardware threads, we would hope to see the speedup improving out to four threads; perhaps not a fourfold speedup but it should be better than the speedup of 1.89 at four threads (1.89 = the ratio of the run time of the sequential program and the parallel program on 4 threads). The topic of characterizing performance is complex and would require an entire book to adequately address. Keeping things simple, the best place to start is to consider the memory movement associated with a computation. In the case of our Pi program, there does not seem to be much occurring in terms of memory traffic. The computation isn't, for example, sweeping through large arrays streaming through memory or moving data to and from a file

---

[3]We only cover the simplest case in this section where the number of threads is a single scalar value. Later in section 12.1.1 we discuss the more complex case where the environment variable defines a hierarchy of threads.

[4]Included with an operating system built around a Linux core (such as OSX) is a command line interpreter or *shell*. Common command line interpreters include Bash, Tcsh/Csh, sh, and Ksh, etc. They overlap considerably so you often do not need to know which shell you are using. They differ, however, in some key aspects, one of which being how environment variables are set. We show how environment variables are set for Bash since it is the most commonly used shell.

system. The only reads and writes to memory are the elements of the array `sum`. The elements of the array are independent. There is no sharing so how could those operations impact scalability?

This is an example of a memory effect called *false sharing*. If a thread accesses a given element of an array, the most likely subsequent memory reference to that array will be the immediately following element. This is called *spatial locality*. To benefit from spatial locality, arrays are decomposed into blocks with each block mapped to an L1 cache line; hence, adjacent array elements tend to travel together in shared L1 cache lines. Information in cache lines can be exchanged between cores as shown in Figure 4.8. In false sharing, when independent data elements happen to be part of the same cache line, each update will cause the cache lines to move back and forth between cores. This excess memory traffic can dramatically slow down a program.

To understand this effect, consider the data movement described in Figure 4.8. Assume an identical copy of the cache line containing Sum[0:3] is located in all L1 caches and HW Thread 0 modifies Sum[0]. This modification invalidates all other copies of this cache line. Unfortunately, this means that HW Thread 2 that is about to update Sum[2] believes the values of Sum[2] it has in its cache line is invalid. While we know this is not true, the system cannot distinguish changes at this level and will copy the freshly modified cache line to the L1 cache of Core$_2$. Hardware thread 2 then updates Sum[2], but this automatically invalidates the other copies of this cache line. Now suppose HW Thread 1 wants to update Sum[1]. It will see an invalid cache line and the system will fetch a copy of the just updated cache line from Core$_2$. This continues with the cache line moving back and forth between the two cores with each update to the Sum. The program is correct. There is no problem with data sharing conflicts between the threads. However, due to the fact that array elements share a cache line, we have a false sharing effect limiting our performance.

One way to remove false sharing is to pad the array responsible for the false sharing. Consider the code in Figure 4.9, we have added a second dimension to the `sum` array declaration in line 13 and made it large enough to fill the L1 cache line (8 doubles). This forces each first element of the array to reside in a different L1 cache line. We waste the space in the cache line and eliminate the possibility of exploiting spatial locality. In this program, however, spatial locality is not important and we are not worried about inefficient utilization of memory since the `sum` array is quite small (even when the padding is taken into account).

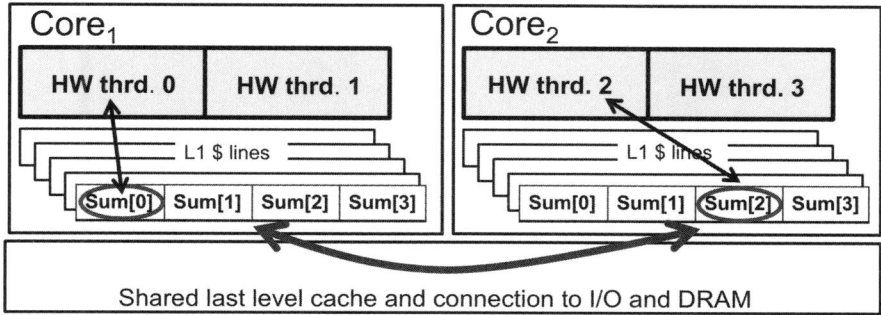

Figure 4.8: **False sharing** – Two cores (1 and 2) and the cache lines used to hold the array Sum. Contiguous elements of the array are often mapped to the same cache line. The data is distinct (there is no true sharing) but cache lines are shared creating high overheads as cache lines move back and forth between the two cores.

The results from the program with a padded sum array are shown in Table 4.8. The performance improvement is striking with improved performance all the way out to 4 threads (a speedup of $1.83/0.53 = 3.4$). Padding the array works well and is a small addition to the program. It requires knowing the size of the L1 cache so it may be necessary to vary the size of the padding dimension as the program moves from one processor to another. There is a second and more subtle problem with padding arrays: It is confusing. In the lifetime of a program, more time is spent maintaining code than writing it. To the reader of a program, perhaps without access to the original author, the addition of a second dimension to the sum array seems arbitrary and confusing.

Table 4.8: **Run times in seconds for the numerical integration program with and without array padding** – Serial program ran in 1.83 seconds. All times are elapsed wallclock times.

| Threads | SPMD | SPMD padded |
|---------|------|-------------|
| 1 | 1.86 | 1.86 |
| 2 | 1.03 | 1.01 |
| 3 | 1.08 | 0.69 |
| 4 | 0.97 | 0.53 |

```
1   #include <stdio.h>
2   #include <omp.h>
3
4   #define NTHREADS 4
5   #define CBLK        8
6
7   static long num_steps = 100000000;
8   double step;
9   int main ()
10  {
11      int i, j, actual_nthreads;
12      double pi, start_time, run_time;
13      double sum[NTHREADS][CBLK]={0.0};
14
15      step = 1.0 / (double) num_steps;
16
17      omp_set_num_threads(NTHREADS);
18
19      start_time = omp_get_wtime();
20      #pragma omp parallel
21      {
22          int i;
23          int id = omp_get_thread_num();
24          int numthreads = omp_get_num_threads();
25          double x;
26
27          if (id == 0) actual_nthreads = numthreads;
28
29          for (i = id; i < num_steps; i += numthreads) {
30              x = (i + 0.5) * step;
31              sum[id][0] += 4.0 / (1.0 + x * x);
32          }
33      } // end of parallel region
34      pi = 0.0;
35      for (i = 0; i < actual_nthreads; i++)
36          pi += sum[i][0];
37
38      pi = step * pi;
39      run_time = omp_get_wtime() - start_time;
40      printf("\n pi is \%f in \%f seconds \%d thrds \n",
41                      pi,run_time,actual_nthreads);
42  }
```

Figure 4.9: **Padded sum array numerical integration** – The sum array padded to fill an L1 cache line with the extra dimension and put subsequent rows of sum, i.e., each sum[id][0], on different cache lines.

The false sharing problem arises from our use of an array for the accumulator. If we could use a scalar declared inside the scope of a structured block executing by a thread, the variables would most likely not reside in the same cache line; false sharing would not be possible. To do this, we need to revisit shared variables and how to use them safely in a program.

### 4.4.3 Synchronization

OpenMP threads execute concurrently; i.e., the instructions from the different threads cannot be placed in a fixed order of execution with respect to each other. There are times when we need concurrent threads to orchestrate their execution so we can constrain the order of certain operations between threads. This orchestration is called *synchronization*. In the OpenMP Common Core, we have two synchronization mechanisms: *critical sections* and *barriers*.

#### 4.4.3.1 Critical

The most basic synchronization construct defines a mutual exclusion relationship for code running with multiple threads. Mutual exclusion stipulates that if one thread is executing a block of code and a second thread tries to execute the same code, that second thread will pause and wait until the first thread has finished with the code. In OpenMP, we define a block of code that executes with mutual exclusion with a `critical` construct as shown in Table 4.9.

Table 4.9: **A *critical construct* defines a block of code that executes with mutual exclusion** – i.e., only one thread at a time executes the code with threads potentially waiting their turn at the beginning of the construct. The curly braces and END CRITICAL are not required when the block contains only a single statement.

```
#pragma omp critical
{
   ...   one or more lines of code
}
!$omp critical
   ...   one or more lines of code
!$omp end critical
```

We provide an example of a `critical` section in Figure 4.10. In this code, we create a team of threads which then execute a loop using a cyclic distribution of loop iterations between threads (as we discussed with the code in Figure 4.6). Inside the loop, the threads execute a statement `B=big_job(i)`. Assume for the sake of this example, that we need to consume the output from `big_job(i)` using a function `consume(B)` and that inside the `consume()` function a number of shared variables are updated. We must assure that those updates are protected so while one thread is doing its updates, another thread does not try to update them at the same time. Otherwise, this would result in a data race and the values of those shared variables inside `consume()` would be undetermined (i.e., there is no way for the system to give them a well-defined value). To prevent data races, we want the `consume()` function to run to completion one thread at a time; that is, we want the function to execute with mutual exclusion. This is precisely what placing the function call inside a critical construct[5] accomplishes.

```
1   #include <stdio.h>
2   #include <omp.h>
3
4   int main()
5   {
6       float  res = 0.0;
7       #pragma omp parallel
8       {
9           float B; int i, id, nthrds;
10          id = omp_get_thread_num();
11          nthrds = omp_get_num_threads();
12          for (i = id; i < niters; i += nthrds) {
13              B = big_job(i);
14              #pragma omp critical
15                  res += consume(B);
16          }
17      } // end of parallel region
18  }
```

Figure 4.10: **An example of a critical section** – Function `consume()` needs to be called by one thread at a time.

Synchronization can be expensive. As we will discuss later when we explore the details of shared memory updates in OpenMP, a synchronization construct implies memory movement to make sure all the threads can see updates from other threads.

---

[5]If the lack of curly braces on the critical construct in Figure 4.10 confuses you, remember that when a structured block has only one line, we can leave off the curly braces, i.e., a single statement is just a block of size one

In most cases, the greatest source of overhead from the use of a critical section emerges from threads waiting their turn to execute the code.

In Figure 4.10 we placed the critical construct inside a loop. In a worst case scenario, the statement B=big_job(i) would execute quickly and the time inside the loop would be consumed by threads waiting at the critical section. In other words, the loop would be serialized. This is disastrous if the goal is good parallel performance. Placing a critical section inside the loop, however, may be OK if the function inside the loop (big_job) runs for a long time relative to the overheads of managing the critical section and if the run time of the function is highly variable from one loop iteration to the next; so that the chances of two threads trying to execute a critical section at the same time are low. In this case we say that the critical section is *uncontended* and the synchronization overhead is manageable.

We can use the critical section to eliminate the need for the sum array in our Pi program. Consider the code in Figure 4.11. We now define sum as a scalar variable and initialize it to zero. Inside the parallel region we define a variable called partial_sum private to each thread. As the loop executes, the sum for each thread is accumulated into this private partial_sum variable. There are no longer any data races since each thread has its own copy of partial_sum. When the loop is done we then add the partial_sum into the sum variable shared between all the threads. This does not create a data race since we put that sum inside a critical section.

Results for the program using the critical section instead of promoting sum to an array or zero padding that array are shown in Table 4.10. Notice how the performance of the code using a critical section closely matches our result from array padding, supporting our assessment that the problem was false sharing. However, just as important as the performance, notice that the code in Figure 4.11 is more portable (i.e., it does not use a constant dependent on the L1 cache size to pad arrays) and is easier to understand.

### 4.4.3.2 Barrier

The most commonly used synchronization construct in OpenMP is a *barrier*. A barrier defines a point in a program at which all threads must arrive before any thread may proceed past the barrier. We have already encountered a barrier when

```
 1  #include <stdio.h>
 2  #include <omp.h>
 3
 4  static long num_steps = 100000000;
 5  double step;
 6  int main ()
 7  {
 8      int i, j, nthreads;
 9      double pi, full_sum = 0.0;
10      double start_time, run_time;
11
12      step = 1.0/(double) num_steps;
13      full_sum = 0.0;
14      start_time = omp_get_wtime();
15  #pragma omp parallel
16  {
17      int i, id = omp_get_thread_num();
18      int numthreads = omp_get_num_threads();
19      double x, partial_sum = 0;
20
21      if (id == 0)
22          nthreads = numthreads;
23
24      for (i = id; i < num_steps; i += numthreads) {
25          x = (i + 0.5) * step;
26          partial_sum += 4.0 / (1.0 + x * x);
27      }
28  #pragma omp critical
29      full_sum += partial_sum;
30  } // end of parallel region
31
32      pi = step * full_sum;
33      run_time = omp_get_wtime() - start_time;
34      printf("\n pi \%f in \%f secs \%d threds \n ",
35          pi, run_time, nthreads);
36  }
```

Figure 4.11: **Numerical integration with a critical section** – The partial sums go into a private variable allocated by each thread. These private variables are extremely unlikely to reside on the same L1 cache lines and therefore, there will be no false sharing. The partial sums are combined inside a critical section so there is no data race.

we discussed the behavior at the end of a parallel construct. When a team of threads reaches the end of the structured block of code within the parallel construct, each thread waits at the end of the structured block until all the threads arrive. Then the master thread of the team, the thread that encountered the parallel construct

**Table 4.10: Run times in seconds for the numerical integration program with and without array padding plus the run time for program using a critical section** – Serial program ran in 1.83 seconds.

| Threads | SPMD no-pad | SPMD padded | SPMD critical |
|:-------:|:-----------:|:-----------:|:-------------:|
| 1 | 1.86 | 1.86 | 1.87 |
| 2 | 1.03 | 1.01 | 1.00 |
| 3 | 1.08 | 0.69 | 0.68 |
| 4 | 0.97 | 0.53 | 0.53 |

in the first place, continues while the other threads shut down. Using the jargon of our synchronization constructs, we say that *the end of a parallel region implies a barrier.*

An explicit barrier can be inserted anywhere within a parallel region using the `barrier` directive, as shown in Table 4.11.

**Table 4.11: An explicit barrier in C/C++ and Fortran** – This defines a point in a program at which all threads in a team must arrive before any threads continue.

```
#pragma omp barrier
!$omp barrier
```

The `barrier` directive is a *stand-alone, executable directive*. It is not associated with a structured block of code and therefore is *not* an OpenMP construct. The `barrier` defines behavior pertaining to how a program executes.

We provide an example of how you might use an explicit barrier in Figure 4.12. This program uses the now familiar SPMD pattern. Each thread calls function `lots_of_work()` to compute a result and places that result in element `Arr[id]` of an array. Notice that later in the program, the array `Arr` is used as an argument to a function. Since we do not know how the array is used inside `needs_all_of_Arr()`, we must assume that any element of `Arr` could be accessed by any thread and therefore, every thread must finish computing its value of `Arr` before proceeding to call `needs_all_of_Arr()`. Hence, an explicit barrier is needed.

A barrier can be very expensive. Any time threads are waiting at a synchronization construct, useful work is not getting done by those threads. This directly translates into parallel overhead. We must use a barrier when it is needed, but a programmer should go to considerable lengths to only use them when the algorithm demands a

```
1   double Arr[8], Brr[8];
2   int numthrds;
3   omp_set_num_threads(8)
4   #pragma omp parallel
5   {
6       int id, nthrds;
7       id = omp_get_thread_num();
8       nthrds = omp_get_num_threads();
9       if (id == 0) numthrds = nthrds;
10      Arr[id] = lots_of_work(id, nthrds);
11  #pragma omp barrier
12      Brr[id] = needs_all_of_Arr(id, nthrds, Arr);
13  } // end of parallel region
```

Figure 4.12: **Example of an explicit barrier** – An explicit barrier is used to assure that all threads complete filling the array **Arr** before using it to compute **Brr**. We assume the SPMD pattern so we pass the thread id and the number of threads to all the functions. Notice that only one thread saves the number of threads to a shared variable should it be needed after the parallel region.

barrier. For example, in Figure 4.12, this program would only be reasonable if the computations inside the function **lots_of_work()** and **needs_all_of_Arr()** ran so long that the overhead associated with the explicit barrier was not significant.

There is one final comment to make about a barrier. In a shared memory computer, a cache coherency protocol makes sure that all threads see a common view of shared memory. The coherency protocol, however, does not describe the detailed timing for when updates to memory appear to other threads. This can get quite complicated, as we will see when we discuss the OpenMP memory model in a later chapter. For the barrier (as well as a critical section), the OpenMP runtime system takes care of these issues on behalf of the programmer. In other words, a thread moving past a synchronization point can assume the OpenMP runtime has done what is required to support a consistent view of memory across the team of threads.

## 4.5   Closing Comments

We have covered a great deal of ground in this chapter. We have covered the core concepts in OpenMP required to write useful multithreaded programs. We have discussed how to create teams of threads (fork) and how they shutdown later on (join). We also talked about the most simple rules for how data is shared between threads and how to create variables that are private to a thread. We described the SPMD pattern which can be used to implement a large variety of parallel algorithms using the number of threads in a team (`omp_get_num_threads()`) and the thread ID (`omp_get_thread_num()`) to divide the work between threads. Looking at a simple numerical integration program, we explored some of the key performance issues in multithreaded programming. This led us into the topic of synchronization and the imposition of ordering constraints between concurrent threads.

You might be surprised at how much you can do with just the little bit of OpenMP that we have covered at this point. There are a host of details on how shared memory works, additional constructs to share work between threads, clauses to manage data that is shared or private, and much more. You can take what we have covered in this chapter, however, and do some serious programming.

# 5 Parallel Loops

Consider a simple program that adds two vectors, a and b.

```
for (i = 0; i < N; i++) {
    a[i] = a[i] + b[i];
}
```

In Chapter 4, we learned how to parallelize loops using the SPMD design pattern. Consider a parallel SPMD version of this vector addition program in Figure 5.1. We create a team of threads using a **parallel** construct. On line 5 we query the thread ID and on line 6 the number of threads. We then use the thread ID and the number of threads to define starting (istart) and ending (iend) positions for a chunk of loop iterations, one chunk per thread. On line 9 we handle the case where the number of threads might not evenly divide the number of loop iterations, N. This is a handy trick well worth remembering. We just set the end of the chunk of loop iterations to the final loop limit N for the last thread in the team. Finally in the loop at line 10 we run the chunks of iterations, one chunk per thread.

```
1   // OpenMP parallel region and SPMD pattern
2   #pragma omp parallel
3   {
4       int id, i, Nthrds, istart, iend;
5       id = omp_get_thread_num();
6       Nthrds = omp_get_num_threads();
7       istart = id * N / Nthrds;
8       iend = (id + 1) * N / Nthrds;
9       if (id == Nthrds - 1) iend = N;
10      for (i = istart; i < iend; i++) {
11          a[i] = a[i] + b[i];
12      }
13  }
```

Figure 5.1: **SPMD parallel vector add program** – Create a team of threads and assign one chunk of loop iterations to each thread.

This approach works and, as we saw in the previous chapter, is an effective way to write scalable, multithreaded programs. The SPMD pattern is probably the most commonly used design pattern in the history of parallel computing. It is also error-prone. We made numerous changes to our original, vector addition loop. Each added line of program text is another opportunity to introduce an error. We need a simpler way to create a parallel loop. In particular, the transformations shown in Figure 5.1 represent a straightforward pattern we can apply to any number of loops. There must be a way to make a compiler automatically carry out such transformations.

This is indeed the case. We call this *loop-level parallelism*. We show the basic code for the loop-level parallel vector addition program in Figure 5.2. As usual with OpenMP, you must have a parallel region to run a program with multiple threads. The parallel region is created with a **parallel** construct on line 2. Immediately prior to the **for** loop we have a directive:

```
#pragma omp for
```

This directive causes the compiler to produce code similar to that shown in Figure 5.1. The iterations of the loop define the work associated with the loop. This work is shared among the team of threads so they can execute in parallel. We achieved this parallelism, however, with much less effort than was required with the SPMD pattern, with the addition of just a couple of directives.

```
1  // OpenMP parallel region and a worksharing−loop construct
2  #pragma omp parallel
3  {
4      #pragma omp for
5          for (i = 0; i < N; i++) {
6              a[i] = a[i] + b[i];
7          }
8  }
```

Figure 5.2: **Loop-level parallelism for the vector add program** – We create a team of threads and then add a single directive to split up loop iterations among threads.

The directive to split up loop iterations among a team of threads is called a *worksharing-loop construct*. To most OpenMP programmers, this construct is the essence of OpenMP. In this chapter, we will describe this construct, how to work with it, and the most commonly used clauses that modify its behavior.

## 5.1   Worksharing-Loop Construct

OpenMP defines several worksharing constructs. A worksharing construct tells the compiler to split up the work in a construct among a team of threads. The most commonly used worksharing construct is the *worksharing-loop construct*. As demonstrated in Figure 5.2, the construct divides the iterations of a loop among a team of threads.

The basic syntax for the worksharing-loop construct is shown in Table 5.1. Notice the keyword `for` used in C/C++ is different from the keywords `do` and `end do` in Fortran. This is the only keyword difference between C/C++ and Fortran in the names of the OpenMP directives. The loop associated with a worksharing-loop construct immediately follows the directive. It must have the following canonical form:

```
for (init-expr; test-expr; incr-expr)
    structured block
```

Table 5.1: **A basic worksharing-loop construct in C/C++ and Fortran** – The worksharing-loop construct shares the iterations of a loop among a team of threads. The loop is called `for` in C and `DO` in Fortran. Fortran is not block structured, so we need an `END DO` directive. Optional clauses give the programmer more control over the loop construct and include `schedule`, `reduction`, and `nowait`. We will discuss these clauses later in this chapter. Additional clauses define storage attributes of the variables used in the worksharing-loop. We will cover those in Chapter 6.

| |
|---|
| **#pragma omp for** *[clause[[,] clause]...]* <br>     for-loop |
| **!$omp do** *[clause[[,] clause]...]* <br>     do-loop <br> **!$omp end do** *[nowait]* |

The loop control index supported in the OpenMP Common Core is a basic integer type[1]. It is initialized by the *init-expr* which is a basic assignment operation. The *test-expr* is a relational expression using one of the common relational operators such as `<`, `<=`, `>`, or `>=`. Finally, the *incr-expression* uses the familiar increment (`++`), decrement (`--`) operators or an integer expression with addition or subtraction by a fixed constant value.

An example of a worksharing-loop construct is shown in Figure 5.3. First we need to create multiple threads with the `parallel` construct in line 1 since an OpenMP

---

[1]In more advanced OpenMP programming, the loop control variable can be a random-access iterator type (in C++) or even a pointer type. We do not include these cases in the OpenMP Common Core.

*worksharing-loop construct* will only run with multiple threads if it is contained within a parallel region. The `for` construct on line 3 directs the compiler to generate code that splits the work for the loop immediately following the `for` construct among the threads in the team. The result is that each thread will be responsible for one or more chunks of loop iterations and calls the function `NEAT_STUFF(i)` for all the loop iterations it is responsible for.

```
1  #pragma omp parallel
2  {
3      #pragma omp for
4          for (i = N; i >= 0; i = i - 2) {
5              NEAT_STUFF(i);
6          }
7  }
```

Figure 5.3: **An example of a parallel worksharing-loop construct** – Create multiple threads, then split the loop iterations among multiple threads to share the work.

Consider the loop control index `i`. Each thread reads and modifies the value of `i` as it executes its set of loop iterations. If this variable is shared among the threads, the reads and updates would conflict in unpredictable ways resulting in a data race. Consequently, OpenMP requires that the compiler generates code so each thread has its own *private* copy of the loop control index (`i` in this case). This rule, however, only applies to the loop immediately following the worksharing-loop construct. If there are other loops nested inside the worksharing-loop, their indices will not be made private. It is the responsibility of the programmer to manage those loop control indices.

For all worksharing constructs, there is an implicit barrier at the end of the construct. All threads wait at the end of the worksharing-loop construct until the full team of threads working on the construct has finished. In the example shown in Figure 5.3, for example, there is an implicit barrier at line 6. The barrier assures that any variable shared among the threads is available to all the threads in the team after the end of the parallel loop.

## 5.2 Combined Parallel Worksharing-Loop Construct

The following pattern with a pair of OpenMP constructs, one to create the team of threads and the other to split up loop iterations among threads, is very common:

```
#pragma omp parallel
{
    #pragma omp for
        for-loop
}
```

As a convenience, these two directives can be combined into a single directive:

```
#pragma omp parallel for
    for-loop
```

This combined construct reduces the number of changes a programmer must make when incorporating parallel loops into a program. We define the combined parallel worksharing-loop construct in Table 5.2.

## 5.3 Reductions

Consider the program in Figure 5.4. We create an array, A, and then initialize it by a call to a function InitA(). Then, we move through the array and accumulate the values of the array elements by summing them into a single variable, ave. In this case we then divide by the length of the array to compute the average, but this general pattern appears quite often in loops we want to work with. The problem is that in this example, the variable ave defines a loop-carried dependence; that is, the value of ave computed in any given iteration of the loop depends on values produced by earlier iterations. Using the OpenMP worksharing-loop construct, there is no way (short of restructuring the loop body as we did with the SPMD pattern) to resolve this dependence and make the loop iterations independent so they can execute in parallel.

Cases such as that shown in Figure 5.4 are extremely common. They are called *reductions*. To handle this case, we added a reduction clause to OpenMP.

```
reduction(op:list)
```

Table 5.2: **Combined parallel worksharing-loop construct in C/C++ and Fortran** – This construct creates a team of threads and splits up the iterations of the following loop among the team of threads. The allowed clauses are the ones allowed with a *parallel* construct or a worksharing-loop construct.

| Separate constructs | **#pragma omp parallel**<br>{<br>    **#pragma omp for**<br>    for-loop<br>} |
|---|---|
| **Combined construct** | **#pragma omp parallel for**<br>for-loop |
| **Separate constructs** | **!$omp parallel**<br>    **!$omp do**<br>        do-loop<br>    **!$omp end do**<br>**!$omp end parallel** |
| **Combined construct** | **!$omp parallel do**<br>    do-loop<br>**!$omp end parallel do** |

```
1   int i;
2   double ave = 0.0, A[N];
3
4   InitA(A, N);
5
6   for (i = 0; i < N; i++) {
7       ave += A[i];
8   }
9   ave = ave/N;
```

Figure 5.4: **A serial reduction example** – This loop has a loop-carried dependence through the variable *ave* and therefore, the loop cannot be parallelized with a worksharing-loop directive without completely changing the body of the loop (as we did with the SPMD examples in Chapter 4).

where `op` is a basic scalar operator such as `+`, `*`, `-`, `min`, `max`, plus logical and bitwise operators and `list` is a comma separated list of variables in shared memory (i.e., all the threads in the team can see the values of these variables). We show the program in Figure 5.4 parallelized with the aid of a reduction operator in Figure 5.5. OpenMP will create a private copy of the variable ave for each thread. This new private variable is initialized to zero (the identity of the + operator). Each thread then calculates a partial sum of A[i] and updates its local variable ave. Then after the loop, before the threads exit the barrier at the end of the loop, the partial sums of ave are combined with the original value of the global copy of ave to produce the final value.

```
1   int i;
2   double ave = 0.0, A[N];
3
4   InitA(A, N);
5
6   #pragma omp parallel for reduction (+:ave)
7       for (i = 0; i < N; i++) {
8           ave += A[i];
9       }
10  ave = ave/N;
```

Figure 5.5: **An OpenMP reduction** –Each thread has a private copy of the variable ave to use for its loop iterations. At the end of the loop, these values are combined to create the final value of the reduction which is then combined with the globally visible, shared copy of the variable ave.

To effectively use the reduction clause, it is important to understand the specific details of how OpenMP manages a reduction.

reduction(op:var)

For each variable in the list (the *reduction variables*), the system will create a private variable of the same name for each thread. At this point, we distinguish between the *original variable* present before the construct containing the reduction clause, and the new private variables local to each thread. Each thread initializes its private copy of the variable to the identity for the operator(op) indicated in the clause (see Table 5.3). Once the private variable has been created and initialized, the code inside the construct executes as usual. When a thread completes its work, it waits at the barrier at the end of the construct. Before the threads complete the construct and exit the barrier, the local copies of the reduction variables from each thread are combined together using the op from the reduction clause to produce the final reduced value, which is then combined with the original variable using op to produce the final result.

It is important to appreciate that inside the loop, the operations with reduction variables carried out by each thread do not need to use the operator specified in the reduction clause. For example, in Figure 5.5, it is perfectly fine for the body of the loop to include statements such as `ave *= A[i]`. The operator specified in the reduction clause is only used to: (1) initialize the reduction variable, and (2) at the end of the parallel loop when the local copies from each thread are combined into the final value.

There are a few details of reductions that are important to understand. Since the operation implied by the reduction is tightly coupled to how the reduction variables are managed, you can only specify one operator in a reduction clause. Hence, it is allowed to have multiple reduction clauses, but a given variable can only appear in a single reduction clause.

It is extremely important to understand that how an implementation of OpenMP combines the results from the partial sums is not defined in the OpenMP specification. Implementations may use `critical`, a binary tree, or some other schemes. Since floating point operations are not strictly associative, this means the result from a reduction may vary from one run of a program to another[2].

Table 5.3 lists the commonly used reduction operators and the values they imply when initializing reduction variables. An operator used in a reduction clause must be associative. The most commonly used operators are addition (+) and multiplication (*) for which the indicated initialized values are zero and one respectively. The initialization values for `min`(*Largest positive number*) and `max` (*Most negative number*) may be system dependent values.

The reduction is a powerful capability. OpenMP includes user-defined reductions and reductions over array sections. However, we do not include them in the Common Core since they raise a number of complications best addressed after a programmer has mastered reductions at the level of the Common Core.

---

[2]Contrary to popular misconceptions, this is not an error in OpenMP. With very few exceptions, if the variation in a program's result due to changing the order in a reduction matters, the algorithm in question is numerically unstable. Blame the algorithm and the data, not OpenMP.

Table 5.3: **Reduction operators and initial values** – A wide range of associative operators can be used with `reduction`. In this table, we show the ones used in the OpenMP Common Core and the initial values they imply for their reduction variables.

| Operator | Initial Value |
|:--------:|:-------------:|
| + | 0 |
| * | 1 |
| - | 0 |
| min | Largest positive number |
| max | Most negative number |

## 5.4   Loop Schedules

When we use a worksharing-loop construct, it is left to the compiler to choose how to split a loop among the threads. A programmer often has knowledge of an algorithm that goes well beyond what a compiler can deduce. Hence, we added a clause that can be placed on a worksharing-loop construct to provide more control over how iterations are scheduled onto the threads. This is done with the `schedule` clause.

In the OpenMP Common Core, we include the two most commonly used *schedule* clauses: `static` and `dynamic`. The syntax of the `schedule` clause is:

```
schedule(static[, chunk])
schedule(dynamic[, chunk])
```

The optional *chunk* size defines the number of loop iterations that make up the fundamental unit of scheduling. The chunk size can be a literal value known at compile time or an integer expression with a shared variable that is calculated at runtime. It is important, however, that all threads see the same value for the `chunk` size for any given schedule so the compiler schedule is consistent among all the threads in a team.

In the following two subsections, we will describe the static and dynamic schedules: the two schedules supported by the OpenMP Common Core.

### 5.4.1   The Static Schedule

The static schedule on a worksharing-loop constructs directs the compiler to define the schedule for mapping loop iterations onto threads at "compile time". When a chunk size is not provided, the compiler will break the loop iterations into number of chunks equal to the number of threads, and assign one chunk to each thread. Hence, `schedule(static[,chunk])` basically splits the loop as was done manually in Figure 5.1.

If a *chunk* size is specified, then OpenMP will break the loop into consecutive contiguous chunks of iterations of size *chunk*. The chunks are assigned to each thread in a round robin fashion, similar to dealing a deck of cards to the team of threads. For example if there are $M$ chunks and a total of $N$ threads, then thread 0 will get the first chunk, thread 1 will get the second chunk, ... thread $(N-1)$ will get the $N^{th}$ chunk, then it goes back to thread 0 which will get the $(N+1)^{th}$ chunk, and thread 1 will get the $(N+2)^{th}$ chunk, ... and so on, until thread `mod(M,N)-1` receives the last $M^{th}$ chunk.

With the *static* schedule, the compiler generates code to assign loop iterations to threads based on the number of iterations in the loop and the chunk size. Closed form expressions implement the logic to create a well-balanced load (i.e., explicit, static load balancing). Since these decisions are fixed at compile time, the scheduling overhead is reduced and the program could potentially run faster.

Selecting the best chunk size to use with a static schedule can be complicated. In practice, you should try a range of chunk sizes and pick the one that works best. There are advantages to a small chunk size. A small chunk size means the scheduler will have a larger number of chunks to work with. These will be distributed using a round-robin or cyclic distribution. This strategy evenly spreads out variations in per-iteration run times across the team of threads. A larger chunk size, however, works well with the cache hierarchy and increase the chances that data can be reused from cache. Furthermore, the larger chunk size combined with deterministic memory access patterns supports the memory prefetch instructions that a compiler might insert into the code. Hitting the right balance between the small chunks (load balancing) and large chunks (memory locality) can greatly improve performance.

Figure 5.6 contains an example of a parallel loop with a *static* schedule. The compute time of each loop iteration is approximately the same so this loop should work well with a static schedule. We will return to this example later and consider changes to the schedule clause on line 19.

### 5.4.2   The Dynamic Schedule

The *static* schedule works well when the loop iterations take about the same amount of time to run. It is also useful when the loop iterations have predictable run times

```
1   #include <stdio.h>
2   #include <math.h>
3   #include <omp.h>
4
5   #define ITER 100000000
6   void main()
7   {
8       int i;
9       double A[BIG_NUM];
10      for (i = 0; i < ITER; ++i)
11          A[i] = 2.0*i;
12
13      #pragma omp parallel
14      {
15          int i;
16          int id = omp_get_thread_num();
17          double tdata = omp_get_wtime();
18
19          #pragma omp for schedule(static)
20          for (i = 1; i < ITER; i++) // notice i starts from 1 since
21                                     // the denominator below cannot be 0
22              A[i] = A[i] * sqrt(i) / pow(sin(i), tan(i));
23          tdata = omp_get_wtime() - tdata;
24
25          if (id == 0) printf("Time spent is %f sec \n", tdata);
26      }
27  }
```

Figure 5.6: **A worksharing-loop with a static schedule** – In this example, the work is predictable and balanced for each loop index. Using the static schedule is expected to work best in this case.

so they can be ordered from long-running iterations to short-running iterations, thereby supporting a well-balanced worksharing-loop execution for small chunk sizes. The static schedule runs into trouble, however, in two different situations.

First, the amount of work per iteration of the loop may vary widely. For example, adaptive mesh refinement and particle-in-cell algorithms have work per iteration that varies widely based on the density of mesh points or the number of particles in any particular cell.

Second, if the processors in a system run at different speeds, the scheduler has no way to take such differences into account. Some threads will complete their work much faster than others depending on which iterations are assigned to the slow processors and which are assigned to the fast processors.

In both cases, the amount of work per loop iteration is only known at runtime. There is no static schedule that will balance the time spent by each thread during the computation. And since the team of threads is not finished until the slowest thread finishes its work, this load imbalance among the threads results in an overall slowdown for the program. For these situations, you want automatic, dynamic load balancing. This is specified in OpenMP with `schedule(dynamic[,chunk])`.

With the *dynamic* schedule, the loop iterations are decomposed into chunks. The number of loop iterations per chunk is given by the value of the `chunk` parameter in the clause. If a `chunk` parameter is not provided, the default value is 1. Each thread is assigned its first chunk. When a thread finishes the work for its assigned chunk, it checks a queue of chunks waiting to execute, grabs the next chunk, and carries out the associated computation. This continues until all the chunks have been computed. The assignment of chunks to threads is made at runtime, so the system can adapt to sources of variability in the computation.

An example of a worksharing-loop with a *dynamic* schedule is shown in Figure 5.7. The compute time of function `isprime()` varies with the function argument `num`. Later in this section, we will show performance comparisons when altering line 28 to select different schedules.

Since the scheduling for the *dynamic* schedule happens at runtime, the scheduling overhead is higher than for the *static* schedule. For a loop where the time to complete each iteration is highly variable, the greater scheduling overhead with the *dynamic* schedule may pay off given that it will result in better load balancing, especially when the computation time per iteration is large relative to the time for managing the scheduling overhead.

### 5.4.3   Choosing a Schedule

Table 5.4 summarizes the schedule clauses. We cover the syntax of the clauses, their features, and when to use them.

If the programmer does not specify a schedule clause for the worksharing-loop construct, the compiler chooses which schedule to use. The OpenMP specification does not require any particular schedule in this case. It is up to the implementation to select a suitable schedule.

One of the challenges in working with OpenMP is to balance the load across threads. If one thread, for example, finishes its work in one minute while the slowest thread finishes in 20 min, the overall program will be observed to take 20

```
1   #include <stdio.h>
2   #include <math.h>
3   #include <stdbool.h>
4   #include <omp.h>
5
6   #define ITER 50000000
7
8   bool check_prime(int num)
9   {
10      int i;
11      for (i = 2; i <= sqrt(num); i++) {
12          if (num % i == 0)
13              return false;
14      }
15      return true;
16  }
17
18  void main( )
19  {
20      int sum = 0;
21
22      #pragma omp parallel
23      {
24          int i;
25          int id = omp_get_thread_num();
26          double tdata = omp_get_wtime();
27
28          #pragma omp for reduction (+:sum) schedule(dynamic)
29          for (i = 2; i <= ITER ; i++) {
30              if (check_prime(i)) sum++;
31          }
32          tdata = omp_get_wtime() - tdata;
33
34          if (id == 0) printf("Number of prime numbers is %d in
35                               %f sec \n", sum, tdata);
36      }
37  }
```

Figure 5.7: **A worksharing-loop with a dynamic schedule** – In this program, the work per iteration is highly variable. The dynamic schedule should be much better at balancing the load across the team of threads.

minutes to complete. The dynamic schedule provides automatic load balancing, however, its runtime scheduling overhead is much higher than that observed with the static schedule. A compromise approach that often works well is to use the static schedule with a moderate chuck size. The chunks effectively spread out

Table 5.4: **Schedule clause for worksharing-loop constructs** – The schedule clause affects how loop iterations are mapped onto threads. This table summarizes the syntax of these clauses, their features, and when to use them.

| Schedule | Static | Dynamic |
|---|---|---|
| **Syntax** | **schedule (static**[*, chunk*]**)** | **schedule (dynamic**[*, chunk*]**)** |
| **Default** | 1 chunk per thread | *chunk* size $= 1$ |
| **When to use** | Predictable, small variation in work per iteration | Unpredictable, highly variable work per iteration |
| **Features** | Least work at runtime. Scheduling logic set at compile time. | Most work at runtime. Complex scheduling logic at runtime. |

the iteration space among the threads with many chunks assigned to each thread. Statistically speaking if the variation across iterations is not too variable and roughly randomly distributed, you achieve an overall schedule that is not optimal but is often "good enough". What you lose in a non-optimal static schedule is more than compensated for by the lower overhead scheduling decisions.

In practice, programmers experiment with different schedules to find the schedule and chunk size that works best for a given problem running with a given implementation of OpenMP on a specific platform. "Rules of thumb" to guide which schedule to use provide a good starting point, but the best schedule is a complex mix of data reuse from cache, memory prefetch, vectorization, and the needs to balance the load in an algorithm. OpenMP programmers try a range of schedules and chunk sizes to find what works best for each situation. For example, in Table 5.5 we show performance results from the example code shown in Figure 5.6 using the *static* and *dynamic* schedules with a few different *chunk* sizes.

In this example, the `schedule` clause in Figure 5.6 line 19 is replaced by the schedules listed in the first column of Table 5.5. The loop iterations in lines 20-22 have similar run times per iteration. We expect that the static schedule will be best suited. The data in Table 5.5 confirms this expectation with the static schedule run time of 2.84 sec being much faster than the dynamic schedule run time of 14.35 sec. For the different chunk sizes, we see that the default case (one large chunk per thread) has a similar run time to the static schedule using chunk sizes of 1, 8, or 20.

Notice that the case `schedule(static,1)` is effectively the cyclic distribution of loop iterations as shown in Figure 4.6 while `schedule(static)` without a chunk

Table 5.5: **Run times in seconds for the code in Figure 5.6** – We run the program with different schedules and different numbers of threads. We used the GNU compiler version 7.3.0 on an Intel® Xeon™ E5-2698 v3 CPU @ 2.30GHz. We used the compiler optimization level -*O3* with serial execution time of 11.08 sec.

| GNU, -O3 | Threads | | | |
|:---:|:---:|:---:|:---:|:---:|
| **Schedule** | **4** | **8** | **16** | **32** |
| default | 2.86 | 1.60 | 0.90 | 0.69 |
| static | 2.84 | 1.57 | 0.90 | 0.67 |
| static,1 | 2.83 | 1.55 | 0.95 | 0.75 |
| static,8 | 2.93 | 1.61 | 0.95 | 0.73 |
| static,20 | 2.92 | 1.67 | 0.93 | 0.73 |
| dynamic | 14.35 | 12.58 | 9.25 | 7.75 |
| dynamic,8 | 5.81 | 2.96 | 2.33 | 0.86 |
| dynamic,20 | 3.66 | 1.90 | 1.03 | 0.70 |

size is effectively the block distribution as shown in Figure 4.7. In this example, we see little difference between the cyclic or block distribution. The block distribution is more effective for exploiting cache locality and memory prefetch. It also works better than the cyclic distribution with compiler vectorization. The fact we see little difference between these two cases makes sense given the code in question does not benefit from cache reuse (i.e., spatial reuse common to array operations is not relevant in this code) and the specific operations in this case are unlikely to make effective use of a vector unit.

Table 5.6 presents results from the example program from Figure 5.7 using the *static* and *dynamic* schedules with a few different *chunk* sizes.

In this example, the `schedule` clause in Figure 5.7 line 28 is replaced with the schedules listed in the first column of Table 5.6. The loop iterations in lines 29-31 run in widely differing amounts of time due to the behavior of the `check_prime` function. We expect this program will benefit from a dynamic schedule. Consider the column for 8 threads. In these cases, the dynamic schedule is better for small chunk sizes. These advantages go away for larger chunk sizes. Consider the case of `schedule(static,8)`. The performance in this case is similar to the various dynamic schedules. This is consistent with the common practice of using static schedules with a small chunk size to achieve a statistically good load balance with reasonably low scheduling overhead at runtime.

Table 5.6: **Run times in seconds for the code shown in Figure 5.7** – We try different schedules and vary the number of threads. We used the GNU compiler 7.3.0 and an Intel(R) Xeon(R) Haswell processor CPU E5-2698 v3 @ 2.30GHz. We used compiler optimization at level -*O3* with serial execution time of 48.37 sec.

| GNU, -O3 | Threads | | | |
|---|---|---|---|---|
| Schedule | 4 | 8 | 16 | 32 |
| default | 16.45 | 8.75 | 5.02 | 3.40 |
| static | 16.58 | 8.80 | 4.99 | 3.28 |
| static,1 | 24.38 | 12.23 | 6.75 | 3.94 |
| static,8 | 12.26 | 6.72 | 3.88 | 2.28 |
| static,20 | 12.38 | 6.74 | 3.85 | 2.08 |
| dynamic | 12.61 | 6.89 | 4.15 | 3.37 |
| dynamic,8 | 12.59 | 6.77 | 3.80 | 2.29 |
| dynamic,20 | 12.37 | 6.73 | 3.78 | 2.69 |

## 5.5 Implicit Barriers and the Nowait Clause

Worksharing constructs have an implied barrier at the end of the construct. This causes the threads to wait at the end of the construct until all the threads in the team have completed their work. Hence, when the threads execute beyond the end of the construct, they can assume computations inside the worksharing construct have completed and any variables updated inside the construct are available to other threads to use. This behavior is safe (i.e., less error-prone) and supports well-defined semantics for the program.

It is a general goal in OpenMP to make the default behavior of constructs the safer behavior. A barrier, however, is an expensive synchronization operation. Since all threads must arrive at the barrier before any can proceed, any load imbalance or a single slow thread can hold up the entire team of threads and greatly increase parallel overhead.

When the semantics of the program demands a barrier, you need it to be there. When optimizing the performance of a parallel program, however, it is important to only use a barrier when the algorithm demands it and to skip it when it is safe to do so. This is especially the case if you want to scale to a large number of threads.

If you can determine that a barrier is not needed at the end of a worksharing construct, you need a way to disable it. For a worksharing-loop construct, you do this with a `nowait` clause. For C/C++, the `nowait` clause is placed on the worksharing-loop construct:

```
#pragma omp for nowait
```

In Fortran, the `nowait` is placed on the `end do` directive:

```
!$omp for
    ... do-loop
!$omp end do nowait
```

We will illustrate the use of the `nowait` clause in Figure 5.8. In this program, there are three large arrays A, B, C. Inside the parallel region, we first get the thread id, then call function `big_calc1(id)` using the now familiar SPMD design pattern. The results are written into the array A.

```
1    double A[big], B[big], C[big];
2
3    #pragma omp parallel
4    {
5        int id = omp_get_thread_num();
6        A[id] = big_calc1(id);
7
8        #pragma omp barrier
9
10       #pragma omp for
11       for (i = 0; i < N; i++) {
12           C[i] = big_calc3(i,A);
13       }
14
15       #pragma omp for nowait
16       for (i = 0; i < N; i++) {
17           B[i] = big_calc2(C, i);
18       }
19
20       A[id] = big_calc4(id);
21   }
```

Figure 5.8: **Using a nowait clause with worksharing-loops** – In this example, we explore the need for barriers and cases where they can be disabled with a `nowait` clause.

Later in the program, A is used in the call to the function `big_calc3(i,A)` to calculate C. To make sure that every thread has finished updating the value of A, an explicit barrier is needed right after the computation of A.

Looking ahead in the program, we note that in the following loop, the array C is used in the calculation of B. Therefore, we must assure that every thread finishes the update of C before starting the loop where B is computed. This barrier synchronization is achieved through the implicit barrier at the end of the loop at line 13.

Once again, looking ahead in the program, we notice that the array B is not needed again within the parallel region. Therefore, the implicit barrier at the end of the loop at line 18 in unnecessary, and we could turn it off by adding the `nowait` clause in line 15. Any thread that finishes with updating B in line 17 is safe to go ahead and start the calculation of A in line 20.

Following the calculation of A we come to the end of the parallel region. A barrier is implied at that point as well. There is no way to turn off that barrier. This is because when a parallel region ends, all threads must finish their work and execute a join, so only the master thread continues. Therefore, all threads must reach this implicit barrier before the master thread continues, and there is no way to turn off the implicit barrier at the end of a parallel region.

A programmer must be very careful with the `nowait` clause. OpenMP is an explicit API. If you turn off a barrier with a `nowait` when one is actually needed, there is no way for OpenMP to dependably detect your error. Synchronization errors are among the most difficult errors in parallel programming to detect since they may lead to race conditions; that is, nondeterministic bugs where the program output changes depending on the details of how threads are scheduled. Therefore, we recommend leaving the barriers in place and only add `nowait` once you have verified correctness for your program, validating the program again after removing any implied barriers.

## 5.6   Pi Program with Parallel Loop Worksharing

Consider the range of Pi programs we have worked with, in Chapter 4, we parallelized the program with an SPMD pattern with a cyclic distribution of loop iterations (Figure 4.6), the SPMD pattern with block distribution of loop iterations (Figure 4.7), SPMD with padding to remove false sharing (Figure 4.9), and SPMD with a `critical` construct (Figure 4.11). In all these versions of the program, the programmer had to explicitly distribute loop iterations among the threads and implement the reduction (i.e., combine the partial sums into a single global sum).

With the worksharing-loop construct, we can use a simple and elegant approach to parallelize the loop; where we let the compiler manage the distribution of loop iterations among the threads and to carry out the reduction. This version of the program is shown in Figure 5.9.

Notice that with the worksharing-loop, we change a minimal number of lines of codes from the original sequential code shown in Figure 4.5. Mostly, we just added two pragma: line 19 "`#pragma omp parallel`" and line 23 of "`#pragma for reduction(+:sum)`". Other lines of code that we added are for printing the final result and for setting (and "getting") the number of threads.

The program begins by forking a team of threads to execute the *parallel* region. We define a local variable, x, for each thread. We then encounter the worksharing-loop construct. A private copy of the loop iteration counter, i, is created for each thread automatically by the worksharing-loop construct. The variable sum is shared among threads when the worksharing-loop construct is encountered, but due to the reduction clause, `reduction(+:sum)`, each thread is given a private copy of sum to use when computing its partial sum. Then after the worksharing-loop construct, the partial sums from each of the threads are combined into a single global sum which is then combined with the original global copy. It is this shared copy of sum after the parallel region, that is used to calculate Pi.

The last column in Table 5.7 shows the results from using the worksharing-loop program in Figure 5.9. This version of the program had some extra overhead compared to the SPMD versions of the program. This OpenMP overhead comes from the code required to manage the worksharing-loop. The overall performance, however, is quite good given that the worksharing-loop version of the program is much simpler to write, easier to maintain, and closer to the serial code. Furthermore, real applications (as opposed to simple "Pi programs") run much longer than the 1 to 2 seconds, so the impact of this additional overhead would be negligible.

To understand these results, consider the speedup, that is, the ratio of the serial program to the various parallel programs. Ideally, the speedup should be equal to the number of threads (i.e., perfect linear speedup).

Due to false sharing, when sum is promoted to an array in the 1st SPMD program, the speedup is poor. When we pad the sum array to make sure no two consecutive elements are in the same cache line, the performance is much better. We used a critical section to remove false sharing without the awkwardness of padding the array. This approach worked nicely with the critical section yielding results similar to the case with a padded array.

```
1   #include <stdio.h>
2   #include <omp.h>
3
4   #define NTHREADS 4
5
6   static long num_steps = 100000000;
7   double step;
8
9   int main()
10  {
11      double x, pi, sum = 0.0;
12      double start_time, run_time;
13      int i;
14
15      step = 1.0 / (double) num_steps;
16      omp_set_num_threads(NTHREADS);
17      start_time = omp_get_wtime();
18
19      #pragma omp parallel
20      {
21          double x;
22
23          #pragma omp for reduction(+:sum)
24          for (i = 0; i < num_steps; i++) {
25              x = (i + 0.5) * step;
26              sum += 4.0 / (1.0 + x * x);
27          }
28      }
29      pi = step * sum;
30      run_time = omp_get_wtime() - start_time;
31      printf("pi is %f in %f seconds \n", pi, run_time);
32  }
```

Figure 5.9: **Pi program with a worksharing-loop and a reduction** – The
program computes the integral of a function by filling the area under a curve with
rectangles and summing their areas. Loop iterations are divided among threads by the
compiler under direction of the worksharing-loop construct. The reduction creates a private
copy of sum for each thread, initializes it to zero, accumulates partial sums into the sum
variable, and then combines partial sums to generate the global sum.

## 5.7   A Loop-Level Parallelism Strategy

In this section, we consider a strategy programmers often use with loop-level
parallelism and OpenMP. We start with an application dominated by a series of
loops. We assume the starting point is a serial program or perhaps an MPI program

Table 5.7: **Run times in seconds for the numerical integration program with and without array padding, using a critical section, and using parallel worksharing-loop** – The serial program (without OpenMP) ran in 1.83 seconds.

| Threads | 1st SPMD | 1st SPMD Padded | SPMD Critical | Pi Loop |
|---------|----------|-----------------|---------------|---------|
| 1 | 1.86 | 1.86 | 1.87 | 1.91 |
| 2 | 1.03 | 1.01 | 1.00 | 1.02 |
| 3 | 1.08 | 0.69 | 0.68 | 0.80 |
| 4 | 0.97 | 0.53 | 0.53 | 0.68 |

where we are adding OpenMP to exploit parallelism "on a node".

The first step is to find the loops that are computationally intensive so the benefits of parallelism will offset the OpenMP scheduling overhead. For this, we recommend OpenMP profiling tools documented at:

```
https://www.openmp.org/resources/openmp-compilers-tools/
```

Next, you inspect the loops to see if they fundamentally can execute in parallel. In other words, do the loops contain concurrency that can be exploited with parallel execution? In most cases, the loops will include some loop-carried dependencies. We *expose the concurrency* by transforming the loops so they are independent and can execute in any order. This involves finding and exploiting reductions, replacing variables updated with increments into ones computed from loop control indices, privatizing read-only, shared data, and other semantically neutral transformation that make loop iterations independent.

For example, consider the program in Figure 5.10. This program contains a loop with a loop-carried dependence because of the way j is defined in line 5. The j of the next iteration is dependent on j of the current iteration. For this example, we can resolve this dependence by computing j directly as j = 5 + 2 *(i+1). Now j is only dependent on the loop index i, and the loop-carried dependence has been removed.

A useful technique to test if loop iterations are truly independent is to execute the loop in reverse, by swapping the start condition and end condition. If the reverse traversal of the loop produces the same result as the forward traversal, the loop is most likely (although no guarantee) free from loop-carried dependencies.

```
1   // Sequential code with loop dependence
2   int i, j, A[MAX];
3   j = 5;
4   for (i = 0; i < MAX; i++) {
5       j += 2;
6       A[i] = big(j);
7   }
8
9   // Parallel code with loop dependence removed
10
11  int i, A[MAX];
12  #pragma omp parallel for
13      for (i = 0; i < MAX; i++) {
14          int j = 5 + 2*(i+1);
15          A[i] = big(j);
16      }
```

Figure 5.10: **Loop dependence example** –The first loop is sequential and contains a loop-carried dependence. The value of j for a loop index is dependent on the value of j for the previous loop index. In the parallel code in the second loop, the loop-carried dependence has been removed by calculating j from the loop control index.

Once your loops have no more loop-carried dependencies, you can *exploit* the concurrency by adding OpenMP directives: to turn the serial program into a parallel program. In many cases, you can do this by adding the directive:

    #pragma omp parallel for

If the loop includes a reduction, add a reduction clause to the worksharing-loop directive.

As a last step, you should experiment with different numbers of threads and loop schedules to optimize your program. Pay close attention to the implicit barriers and see if they can be safely be turned off with a nowait clause. Only experiment with nowait clauses after you have an effective testing regime so you can validate you did not introduce any race conditions when removing a barrier.

## 5.8   Closing Comments

Parallel loops are the heart and soul of OpenMP. Most OpenMP programmers rarely move beyond the basic parallel worksharing-loop combined construct. They find the most time-consuming loops, transform them so loop iterations are independent, add the parallel worksharing-loop directive, and call the parallelization effort "done".

We urge programmers, however, to go beyond the trivial worksharing-loop constructs. Time spent thinking about load balancing and other performance issues can pay off nicely and deliver much better speedups. Programmers should spend time experimenting with different loop schedules. Consider the data structures in the body of a loop and how those structures map onto the caches in the multiprocessor system. This information might suggest chunk sizes for the schedule clauses which could dramatically improve performance.

Performance is the goal, but it is important to use a disciplined strategy where you convert a serial program into a parallel program through a sequence of distinct transformations. After each transformation, test the program and take the time to convince yourself the program is correct at each stage of development. This can pay off dramatically as you debug your parallel loops.

# 6 OpenMP Data Environment

Let's review our progress in this journey through the OpenMP Common Core. We started with threads: how to create them and how to use them. A thread is an execution entity. It executes statements from a program and in doing so modifies items stored in memory. An item resides in memory at a specified address. We assign a name to this address and call it a *variable*. In other words, memory is just a collection of variables visible to different elements of a program. Variables are names for addresses so we often describe memory as an *address space*.

An OpenMP program uses a parallel construct to create a team of threads. All the code executed from within a parallel construct is called a *parallel region*. Using just a parallel construct with all the threads executing the same code, we can accomplish a great deal. These algorithms are said to use the *SPMD Pattern*. Alternatively, we can use a worksharing-loop construct inside a parallel region to distribute work among the team of threads. All the code executed inside the worksharing-loop defines a worksharing-loop region. Algorithms defined around the worksharing-loop construct use the *loop-level parallelism pattern*. With both the SPMD pattern and the loop-level parallelism pattern, the threads execute code as they work within a region (the parallel region or the worksharing-loop region). We call the set of variables visible to a thread as it executes a region the *data environment* for that region.

An OpenMP thread has access to a *shared* address space and an address space *private* to the thread. Hence, variables in the OpenMP Common Core can be assigned one of two storage attributes: *private* or *shared*[1]. If a variable can be accessed by one thread and the other threads in the team have no way to see that variable, the variable is *private* or equivalently *local* to a thread. When all the threads in a team can access the variable (i.e., both read and write that variable), we say that the variable is *shared* or *global*.

Up to this point in the book, we have said that if a variable is declared prior to a parallel region, it will be shared among the threads in a team. If a variable is declared inside the structured block associated with a parallel construct, it is private to each thread. This simple pair of rules governs much of what we do with OpenMP and is sufficient for a wide range of parallel algorithms.

---

[1]There is a third storage attribute in OpenMP that is not included in the Common Core. It is called *threadprivate* and will be covered in Section 10.2.1.

There are exceptions, however, and corner cases where the default rules go a bit further. Furthermore, there are times when a programmer needs to change the data environment and directly modify the storage attributes of variables. In this chapter, we address the full range of issues pertaining to the OpenMP Common Core data environment. We will start by exploring the default rules in full detail, so you can assign the right storage attribute to any variable in a program. Then we will describe a set of clauses used in the OpenMP Common Core to modify the data environment.

## 6.1   Default Storage Attributes

OpenMP is a shared memory programming model. Most variables are shared by default. To understand the implications of shared memory, we need to understand how an operating system views a thread.

When a program is launched, the operating system creates a process to run the program. The process contains one or more threads, a block of memory visible to the threads (usually managed as a heap), and any other resources needed to interact with the system. Variables start in the heap memory that the process manages. They are visible to all the threads and thus, are said to be shared.

A process creates (or *forks*) threads. Each thread is created with its own program counter, and its own block of memory (usually managed as a stack) local to a thread. Variables in the thread's stack are only seen by that thread, hence they are private to the thread.

We will now consider default storage attributes for variables in OpenMP and common exceptions. The most basic rule is that variables declared in memory that is visible across all the threads are shared. This memory is the heap managed by the process. A variable declared prior to a parallel region is placed in that process-memory and is shared.

Variables of global scope in the base languages are shared among threads. For example, file scope variables in C/C++ are shared. These variable are declared prior to the main program and are therefore visible to all functions declared in that compilation unit (hence why they are called "file scope"). Variables in C/C++ that are declared with a static storage class are shared. In Fortran, variables in a common block, save variables, and module variables wherever they are declared are shared. Heap variables in dynamically allocated memory in Fortran (via allocate) and C/C++ (via malloc or new) that belong to the process which contains the team of threads are shared.

Not all variables, however, are shared by default. If a variable is on a thread's stack, it is private to that thread. Stack variables in Fortran[2] subprograms or C/C++ functions called from parallel regions are private. Variables declared within a statement block or inside the body of a function are automatic variables. These are on the stack memory, so they are private. Loop indices on a worksharing-loop construct are also private. If a variable is declared inside the parallel region, it is local to each thread, so it is private.

A simple way to think about it is that variables allocated in the process heap are shared, and variables on a thread's stack are private.

In Figure 6.1 we provide an example that illustrates the rules for the default storage attributes.

```
1   // File #1:
2   double A[10];
3   int main()
4   {
5       int index[10];
6       #pragma omp parallel
7           work(index);
8       printf("%d\n", index[0]);
9   }
10
11  // File #2:
12  extern double A[10];
13  void work(int *index) {
14      double temp[10];
15      static int count;
16      ...
17  }
```

Figure 6.1: **An example of default storage attributes** – A, index, count are shared variables since A is a file scope variable, index is defined prior to the parallel region, and count is a static variable. temp is a private variable since it is declared inside the parallel region.

---

[2]Fortran does not require that subprogram variables are managed on a stack unless the subprograms are specified as RECURSIVE. OpenMP compilers usually imply RECURSIVE.

In this example, we see that before the parallel region has created any threads, the variables `A`, `index`, and `count` are placed in the process's memory. According to the rules for the storage attributes of variables in OpenMP, these are therefore shared among threads created by this process. The array `A` is a file scope variable defined before the main program, so it has a global scope. `index` is declared inside the main program but prior to the parallel region, it is visible to any threads created by the process. The variable `count` is declared as `static`, so its storage class is explicitly declared as static making it a shared variable. All the threads can see variables `A`, `index`, and `count`. They are shared.

Inside the main program, an array `index` of dimension 10 is declared. We create a parallel region and call a function `work()` passing `index` as argument. In File #2, `A` is extern to function `work`, meaning the function expects to find an external file scope variable `A[10]`. Inside the function `work()`, `index` is passed as a pointer to the reference of the array. Then we declare a `temp` array and a static variable `count` inside this function.

Since `temp` is declared inside the function `work()`, it resides in the function's stack and is stored in the stack memory of each thread. Each thread will therefore have its own private copy of `temp`.

When running this program with 3 threads, the variables visible to each thread are shown in Figure 6.2. All 3 threads see `A`, `index`, and `count` inside the parallel region since they are shared variables. The variable `temp` is private to each thread, that is, each thread has its own copy of `temp`. At the end of the function `work()`, `temp` goes out of the scope. Hence after the parallel region exits, only `A`, `index`, and `count` are visible to the master thread.

## 6.2   Modifying Storage Attributes

There are circumstances where we need to modify the storage attributes of variables. We do this with clauses placed on directives for the `parallel` or `worksharing-loop` constructs. In the OpenMP Common Core we include the clauses `shared`, `private`, and `firstprivate`. Each of these clauses include a comma-separated list of variables.

To understand these clauses, consider how they appear to a thread executing code in a region. The thread is executing a sequence of instructions when it *encounters* an OpenMP construct. This thread (the "encountering thread") is working within a region with its own data environment and therefore has a set of variables visible to it. The *encountering thread* creates a new region based on the construct it encountered (for example, creating a new parallel region). The new region implies a new data environment.

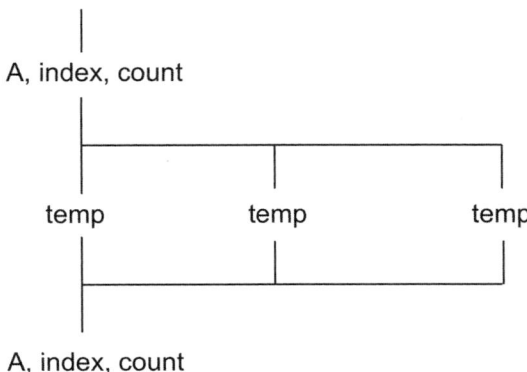

Figure 6.2: **Illustration for the data sharing example in Figure 6.1 with 3 threads.**

Data environment clauses modify the storage attributes for variables in the data environment of the encountering thread that map into the data environment created by the construct. Therefore, there is a single name used for both the variable prior to the construct and inside the construct. When we use that name to refer to the variable in the data environment of the encountering thread (i.e., prior to the construct), we call it the *original variable*. If this concept seems unduly convoluted, it will become clear as we consider some examples in the next few sections.

### 6.2.1   The Shared Clause

A thread encounters an OpenMP construct and creates a new region. A region has an associated data environment. The data environment clauses map variables in the data environment of the encountering thread onto the data environment of the new, just created region. The clauses set the data sharing attributes of variables as they map into the new region.

The default data sharing attribute for these variables is *shared*. If a variable is shared, it means that there is one copy of a variable and it is visible to all the threads in the team. The syntax of the shared clause is:

```
shared(list)
```

where `list` is a comma separated list of variables. We provide an example of a parallel region with a `shared` clause in Figure 6.3. In this example, we declare three integers A, B, and C and initialize them (with an **extern** function we do not show). The thread started with the program (the so-called *initial thread*) encounters the **parallel** construct. It has a single clause:

> #pragma omp parallel shared(B,C)

The clause indicates that the variables B and C are shared inside the parallel region. Notice that this is the default behavior. In fact the variable A is shared as well except we mask the original variable A by creating a new variable of the same name inside the parallel region. Each thread executing the parallel region will create that variable A which will be private to each thread.

```
1   #include <omp.h>
2   #include <stdio.h>
3
4   extern void initialize(int*, int*, int*);
5
6   int main ()
7   {
8       int A, B, C;
9
10      initialize(&A, &B, &C);
11
12      //remember the value of A before the parallel region
13      printf("A before = %d\n", A);
14
15      #pragma omp parallel shared(B,C)
16      {
17          int A = omp_get_thread_num();
18          #pragma omp critical
19              C = B + A;
20      }
21
22      // A in the parallel region goes out of scope, we revert
23      // to the original variable for A
24      printf("A after = %d and C = %d\n", A, C);
25   }
```

Figure 6.3: **The shared clause** – An example of a shared clause on a parallel construct. Strictly this clause is not needed. It is included in this case to remind the programmer that of the three variables A, B, and C, only B and C are shared since A is masked by its declaration inside the parallel region.

Of course, shared variables are visible to all threads. It is essential, to make sure there are no data races that occur as multiple threads try to update the same shared variable at the same time. Hence, when threads write to the variable C inside the parallel region, we protect those updates with a critical section.

At the end of the parallel region, the private copy of A goes out of scope and the variable no longer exists. The name A once again points to the original variable. Hence the two print statements for A in Figure 6.3 will report the same value for A.

The shared(B,C) clause is not needed here. The program would generate the same result with and without the shared clause. Although not required, it is good programming practice to list the shared variables with the shared clause for debugging and to aid in reading the program. This is an important point. One of the leading causes of errors in OpenMP programs is to lose track of the storage attributes of variables inside different regions. A simple reminder of which variables are shared can payoff tremendously for the programmer trying to debug an OpenMP code.

### 6.2.2   The Private Clause

The *private* clause is placed on a directive that generates a new data environment. The syntax for the clause follows:

    private(list)

where list is a comma-separated list of variables defined in the data environment of the thread that encounters the construct with the private clause. The private clause changes the default behavior when mapping variables into the new data environment. The private clause tells the compiler to create a new variable of the same type and name for each thread[3]. This new variable is private to each thread; that is each thread has a copy of the variable that the other threads cannot access.

The value of the private variable is uninitialized. It masks the original variable inside the parallel region, so the thread only sees the private variable. When the region is over, the private variable goes out of scope. This is C/C++ language terminology to say the variable no longer exists after the region that contained the variable completes. Moving beyond the region, the name of the variable refers to the original variable whose value is unchanged by the private variable.

---

[3]In C++ the default constructor is used to create the private variable.

Figure 6.4 shows an example of an OpenMP construct with a *private* clause. Based on our discussion of the private clause, is this program correct?

```
1   void wrong()
2   {
3       int tmp = 0;
4       #pragma omp parallel for private(tmp)
5           for (int j = 0; j < 1000; j++)
6               tmp += j;
7           printf("%d\n", tmp);   //tmp is 0 here
8   }
```

Figure 6.4: **An example of a private clause** – The original variable `tmp` is masked by the private copy of the variable inside the parallel for region. Question: Is this program correct?

The code in Figure 6.4 is incorrect. Since `tmp` inside the parallel for construct was created by a private clause, each thread has a local copy that is not initialized. Hence, the update of `tmp` on line 6 is incorrect since `tmp` does not have an initial value. At the end of the parallel for construct, however, when tmp is printed, it would print the value from its original variable, which in this case is 0. This is because the private copies of `tmp` have gone out of scope after exiting the parallel for construct, so the name `tmp` reverts to the original variable.

In Figure 6.5, we show another example that uses the private clause. This example too has a serious problem. The figure includes two program fragments. Assume these are spread between two files. In File #1, `tmp` is declared as a file scope variable. Then inside the function `danger()`, it is initialized to 0. When the parallel region inside `danger()` is created, the variable `tmp` is privatized.

```
#pragma omp parallel private(tmp)
```

The private variable `tmp` inside the parallel region masks the original variable, which is the original file scope variable with the value of 0. Inside the parallel region, the function `work` defined in File #2 is called. The function `work` expects an extern global scope variable `tmp`, and updates the value `tmp` inside the function. This program is semantically ambiguous. If the function is left in a separate file, the compiler is likely to select the file scope copy of `tmp`. If the compiler inlines the function `work()` it would select the private copy of `tmp`. The issue is that the compiler has no way to figure out which copy of `tmp` it needs to use inside `work()`.

```
1   // File #1
2   int tmp;
3   void danger()
4   {
5       tmp = 0;
6       #pragma omp parallel private(tmp)
7           work();
8       printf("%d\n", tmp);    // tmp has unspecified value
9   }
10
11  // File #2
12  extern int tmp;
13  void work()
14  {
15      tmp = 5;
16  }
```

Figure 6.5: **A second example of the private clause** – This program has a subtle bug. Compilers cannot decide the scope of `tmp`, whether the file scope variable, or the private copy inside the parallel region. Hence, this case is specifically forbidden in the OpenMP specification.

The solution is to specifically define this as an invalid program. The program is ambiguous and therefore the safest course is for the OpenMP specification to disallow the case of a variable being both private and of file scope within a data environment. You cannot expect a compiler to warn you of this situation, since the different treatments of the variable `tmp` are in two different source files. Hence, it is up to the programmer to avoid this case. Fortunately, common programming practice is to avoid file scope variables so this problem does not come up very often.

### 6.2.3   The Firstprivate Clause

The *firstprivate* clause is placed on a directive that generates a new data environment. The syntax for the clause follows:

```
firstprivate(list)
```

where `list` is a comma-separated list of variables defined in the data environment of the thread that encounters the construct with the `firstprivate` clause. Similar to the `private` clause, the `firstprivate` clause tells the compiler to create a private variable for each thread with the same name as the original variable for each item in the list[4]. The *firstprivate* variable also masks the original variable inside the parallel region, and the value of the original variable is unchanged after the region.

The difference between `firstprivate` and `private` is that with `firstprivate`, the new private variable is initialized by copying the value of the corresponding original variable.

In Figure 6.6 we provide an example of the *firstprivate* clause. `incr` is initialized to 0 before the parallel region. With `incr` as a firstprivate variable, each thread has its local copy of `incr` with the initial value of 0 from the original variable. Other than the fact they are initialized, the variables created by a `firstprivate` clause are identical to those created with a `private` clause.

```
1  incr = 0;
2  #pragma omp parallel for firstprivate(incr)
3  for (i = 0; i <= MAX; i++)
4  {
5      if ( (i % 2) == 0) incr++;
6      A[i] = incr;
7  }
```

Figure 6.6: **Example of using the firstprivate clause** – `incr` is a firstprivate variable so it is private to each thread and has an initial value (zero).

### 6.2.4   The Default Clause

In the OpenMP Common Core, we have three clauses to modify the data environment generated by an OpenMP construct: `private`, `firstprivate`, and `shared`. One of the most common sources of errors in an OpenMP program is to have the wrong storage attribute on a variable. It is surprisingly easy to create such errors given some variables import their data sharing attribute by default and others through explicit clauses.

OpenMP provides a `default` clause. In the Common Core, we support one case with the default clause: the `default(none)` clause. If `default(none)` is used on a construct, then all variables passing from the encountering thread into a region must be explicitly listed in a `private`, `firstprivate`, `reduction`, or `shared` clause. The

---

[4]In C++, the copy constructor is used to copy the value from the original variable into the private variable.

compiler will flag any variables not listed in one of the data environment clauses as an error. This can help greatly when trying to understand the variables mapping into a data region and their data sharing attributes.

There are other cases supported for `default` clause such as `default(shared)` or `default(private)` (not available in C/C++). We do not support these additional cases in the Common Core and frankly, they are not used very often.

## 6.3 Data Environment Examples

The best way to move information from short term memory into long term memory is to solve problems. In that spirit, we provide two problems that require an understanding of the data environment clauses. We urge you to look at a problem and then stop reading while you solve the problem. Only after you have your own solution, read our discussion of the problem.

### 6.3.1 A Data Scope Test

Figure 6.7 shows a program in which a thread encounters a parallel construct. We define three variables A, B, and C which are all initialized to 1. We then create a parallel region and specify B as private and C as firstprivate.

```
1        A = 1
2        B = 1
3        C = 1
4        #pragma omp parallel private(B) firstprivate(C)
```

Figure 6.7: **An OpenMP data environment quiz** – Consider the storage attributes and values for A, B and C.

Consider the following questions:

- Specify the storage attribute of A, B, and C inside the parallel region.

- What are their initial values inside the parallel region?

- What are their values after the parallel region?

It is important that you try to answer this question before reading any further.

A was not declared in any clause, so it is a shared variable by default. Every thread sees its value of 1 when entering the parallel region.

The storage attributes of B and C are modified with the *private* and *firstprivate* clauses, so they are both private to each thread. Inside the parallel region, each thread has a local copy of B and C. The value of B for each thread is uninitialized, and the value of C is initialized to the value of 1 from its original variable C.

Following the parallel region, the private copies of B and C go out of scope and B and C revert to the global values of their original variables. Therefore, both have their original value of 1. Since A is shared, it will have whatever value was set inside the parallel region.

You may wonder if there are benefits to using *private* instead of *firstprivate*, since it seems that *firstprivate* achieves everything *private* does but with the safety of a well-defined initial value. Our recommendation is that you should mostly use *private* and only use *firstprivate* when you need an initialized private variable. The *firstprivate* clause can add considerable overhead in copying the original variable's value into the newly created private variable. It may not matter much for simple scalar variables, but variables in data environment clauses can be arrays or objects with complicated data structures. They can imply the copying of large amounts of data with memory operations from a large number of threads.

Another point of discussion is whether we should just declare a private variable inside a construct instead of through a `private` clause. Both approaches have their strengths and weaknesses. It largely comes down to personal style.

One approach focuses on code with identical semantics when interpreted with or without OpenMP. Remember, if a compiler does not recognize a directive, it skips the directive. Hence if programmers are careful, they can write code that works regardless of whether a compiler understands OpenMP. This approach is supported by declaring private variables inside the parallel regions instead of using the *private* clause.

Another approach is to add as few lines of code to a program as possible. Every line of code added to a program creates another opportunity for introducing an error in the code. Therefore we should use OpenMP clauses as much as possible and avoid changes in the original source code.

Both approaches are correct. The choice is a matter of programmer style.

### 6.3.2   Mandelbrot Set Area

In this section, we consider a long and involved exercise to test your knowledge of the data environment clauses. In Figures 6.8 and 6.9 we provide a program that computes the area of a Mandelbrot set. The Mandelbrot set is the set of complex numbers c for which the function $z^2$ + c does not diverge when iterated from z = 0. The area of the Mandelbrot set is known to be around 1.506.

In lines 27-34, with the combined OpenMP parallel for constructs, the program loops over a grid of points in the complex plane which contains the Mandelbrot set, and tests each point to see whether it is inside or outside the set. The actual work of testing a point is in a separate function `testpoint()`.

This program has multiple bugs. Most (but not all) are related to the data environment. We know the program has bugs since it gives different and incorrect results every time you run the program. We urge you to study the program and see if you can find all the bugs. Only after you are convinced you have the right answer should you read any further in this text[5].

The first problem we notice is that `eps` is specified as *private*. Therefore, it has not been initialized and has no value when entering the parallel region even though it is used when updating the value of c. An easy solution is to change the storage attribute for `eps` to *firstprivate*. This gives each thread its own copy of the variable but with a specified value. Notice that `eps` is read-only. It is not updated inside the parallel region. Therefore, another solution is to let it be shared (`shared(eps)`) or not specify `eps` in a data environment clause and let its default, shared behavior be used. While this would result in the correct code, it would potentially increase overhead. If `eps` is shared, every thread will be reading the same address in memory. When the read-only data structure is complicated, and the memory size is large, reading from each thread will be slower from the shared memory than from its own local private memory with the copy of that read-only variable. Some compilers will optimize for such read-only variables by putting them into registers, but we should not rely on that behavior.

To check the data storage attributes for other variables used in the parallel region, we use the *default(none)* clause, which will force the programmer to declare every variable used inside the parallel region. For this problem, when we add *default(none)*, the compiler immediately points out that variable j does not have its data sharing

---

[5]The Mandelbrot program is available from the OpenMP Common Core web site, `http://www.ompcore.com`

```
1   #include <stdio.h>
2   #include <stdlib.h>
3   #include <math.h>
4   #include <omp.h>
5
6   # define NPOINTS 1000
7   # define MAXITER 1000
8
9   void testpoint(void);
10
11  struct d_complex {
12      double r;
13      double i;
14  };
15
16  struct d_complex c;
17  int numoutside = 0;
18
19  int main() {
20      int i, j;
21      double area, error, eps = 1.0e-5;
22
23  // Loop over grid of points in the complex plane which contains
24  // the Mandelbrot set, test each point to see whether it is
25  // inside or outside the set
26
27  #pragma omp parallel for private(c,eps)
28      for (i = 0; i < NPOINTS; i++) {
29          for (j = 0; j < NPOINTS; j++) {
30              c.r = -2.0 + 2.5 * (double)(i) / (double)(NPOINTS) + eps;
31              c.i = 1.125 * (double)(j) / (double)(NPOINTS) + eps;
32              testpoint();
33          }
34      }
35
36  // Calculate area of set and error estimate and output the results
37
38      area = 2.0 * 2.5 * 1.125 * (double)(NPOINTS * NPOINTS - numoutside)
39             / (double)(NPOINTS * NPOINTS);
40      error = area / (double)NPOINTS;
41
42      printf("Area of Mandlebrot set = %12.8f +/- %12.8f\n",area,error);
43      printf("Correct answer should be around 1.506\n");
44  }
```

Figure 6.8: **Mandelbrot set area: original wrong code, part 1** – This version of the program has multiple bugs. Your job is to inspect the code and find the bugs.

```
1    void testpoint(void) {
2
3    // Does the iteration z=z*z+c, until |z| > 2 when point is known to
4    // be outside set. If loop count reaches MAXITER, point is considered
5    // to be inside the set.
6
7        struct d_complex z;
8        int iter;
9        double temp;
10
11       z = c;
12       for (iter = 0; iter < MAXITER; iter++) {
13           temp = (z.r * z.r) - (z.i * z.i) + c.r;
14           z.i = z.r * z.i * 2 + c.i;
15           z.r = temp;
16           if ((z.r * z.r + z.i * z.i) > 4.0) {
17               numoutside++;
18               break;
19           }
20       }
21   }
```

Figure 6.9: **Mandelbrot set area original wrong code, part 2** – This version of the program has multiple bugs. Your job is to inspect the code and find the bugs.

attribute defined. This is the loop control index for the loop nested inside the worksharing-loop. It is not made private by default (as is done for i and the worksharing-loop), so we need to declare j as *private* explicitly[6].

The next problem we notice is that the function `testpoint()` works with a file scope variable for c, but this same variable is specified as a private variable in the parallel worksharing-loop region. As we discussed in the example in Figure 6.5, there is no unambiguous way for the compiler to tell `testpoint` which c should be used. This is an invalid program since we have a single variable that is used as a private variable and as a file scope variable. To fix this problem, we modify the argument list to `testpoint()` so we pass c into the function as an argument. We then remove the file scope copy of the variable since it is no longer used. We then make c *private*, so that each thread has its own copy.

If you were to make these changes and run the program, the answer would still

---

[6]Another solution is to use the more modern C idiom of declaring the loop index where it is used, namely inside the loop, `for(int j=0; j< NPOINTS; j++)`

change from one run to the next. This suggests that there is a data race somewhere in the program. If you look at the increment of the shared variable `numoutside`, it is happening from multiple threads without any constructs to enforce mutual exclusion. That increment operation needs to be protected with a *critical* section; thereby eliminating the data race.

Synchronization is expensive. There are two situations that arise with mutual-exclusion synchronization (such as `critical`, *contended* and *uncontended*). Contended synchronization means that when one thread encounters a synchronization construct, there is a high probability that there are other threads already waiting at that construct, therefore forcing the thread to wait until the other threads complete. The opposite case is when a synchronization construct is uncontended. This means that the chances are high that only one thread at a time will encounter the synchronization construct.

In this Mandelbrot set area example, it is likely we have uncontended synchronization. This is because the times spent by each thread on different iterations are likely quite different. Some points require many iterations to test convergence and others take just a few. Therefore, it is likely that each thread arrives at the critical section at different times. Combined with the fact that the update of `numoutside` is so quick, the chance of contended synchronization is quite low.

In summary, we have identified a total of 4 issues. The errors in the original buggy Mandelbrot sea code are:

1. `eps` was not initialized.

2. The loop index `j` needed to be made *private*.

3. Updates of `numoutside` must be protected with a *critical* construct.

4. The variable `c` must not be a file scope variable. It needs to be private for each thread.

We show the correct version of the program in Figures 6.10 and 6.11. When we run this program multiple times with different numbers of threads, we get correct and consistent results each time.

```
1   #include <omp.h>
2   # define NPOINTS 1000
3   # define MXITR 1000
4   struct d_complex {
5       double r; double i;
6   };
7
8   void testpoint(struct d_complex);
9   struct d_complex c;
10  int numoutside = 0;
11
12  int main ()
13  {
14      int i, j;
15      double area, error, eps = 1.0e−5;
16      #pragma omp parallel for private(c,j) firstprivate(eps)
17          for (i = 0; i < NPOINTS; i++) {
18              for (j = 0; j < NPOINTS; j++) {
19                  c.r = −2.0 + 2.5 * (double)(i)/(double)(NPOINTS) + eps;
20                  c.i = 1.125 * (double)(j)/(double)(NPOINTS) + eps;
21                  testpoint(c);
22              }
23          }
24      area = 2.0 * 2.5 * 1.125 * (double)(NPOINTS * NPOINTS        \
25              − numoutside)/(double)(NPOINTS * NPOINTS);
26      error = area / (double)NPOINTS;
27  }
```

Figure 6.10: **Mandelbrot set area solution, part 1** − c is passed as an argument to function testpoint. eps is declared as firstprivate and the inner loop index j is declared as private.

### 6.3.3  Pi Loop Example Revisited

In the Pi program in Figure 5.9, we use a worksharing-loop construct to carry out the numerical integration. We needed the declaration double x inside the parallel region so each thread would have its own value for the location of each rectangle in the sum. In Chapter 5, we had not yet covered the data environment clauses so we had no choice but to use an explicit declaration.

In Figure 6.12 we provide a simpler implementation of the Pi program. This version minimizes the number of changes needed to convert the serial code to an OpenMP parallel program. A single pragma is sufficient to parallelize the program:

```
#pragma omp parallel for private(x) reduction(+:sum)
```

```
1   void testpoint(struct d_complex c)
2   {
3       struct d_complex z;
4       int iter;
5       double temp;
6
7       z = c;
8       for (iter = 0; iter < MXITR; iter++) {
9           temp = (z.r * z.r) - (z.i * z.i) + c.r;
10          z.i = z.r * z.i * 2 + c.i;
11          z.r = temp;
12          if ((z.r * z.r + z.i * z.i) > 4.0) {
13              #pragma omp critical
14                  numoutside++;
15              break;
16          }
17      }
18  }
```

Figure 6.11: **Mandelbrot set area solution, part 2** – c is passed as an argument to function `testpoint`. The update to `numoutside` is protected with a critical section.

This program runs correctly when compiled with a compiler that understands OpenMP. It also works with a compiler that does not support OpenMP since the `#pragma omp` directive will be ignored in that case. Using the *private* clause supports an elegant solution to our Pi problem while following the common OpenMP design goal of not breaking or changing the serial program when parallelizing it.

## 6.4   Arrays and Pointers

Our focus in this chapter has been on scalar variables. We would be remiss, however, if we left you with the impression that you could only put scalar variables in the data environment clauses. Pointers and arrays can be used inside data environment clauses. We will consider two basic cases. To understand these cases, "think like a compiler".

**static arrays**: Arrays that are declared at compile time with a known size are allocated on the stack. The compiler knows the size of this array and can manage it for you. Consider the code in Figure 6.13. The compiler knows the type and size of the array. It can create a private copy of the array for each thread. If we used

```
1   #include <stdio.h>
2   #include <omp.h>
3
4   #define NTHREADS 4
5
6   static long num_steps = 100000000;
7   double step;
8
9   int main ()
10  {
11      int i;
12      double x, pi, sum = 0.0;
13      double start_time, run_time;
14
15      step = 1.0/(double) num_steps;
16      omp_set_num_threads(NTHREADS);
17      start_time = omp_get_wtime();
18
19      #pragma omp parallel for private(x) reduction(+:sum)
20          for (i = 0; i < num_steps; i++) {
21              x = (i + 0.5) * step;
22              sum += 4.0 / (1.0 + x * x);
23          }
24
25      pi = step * sum;
26      run_time = omp_get_wtime() - start_time;
27      printf("pi is %f in %f seconds %d threads\n", pi, run_time);
28  }
```

Figure 6.12: **Pi Program with combined parallel worksharing-loop and reduction** – Each thread accumulates its local sum that is later combined into the global sum with the reduction operation. Variable x is declared as private with a data environment clause.

firstprivate instead, it would have copied the values into the new private array. We can also use a static array in a reduction clause.

```
int varray[1000];
initv(1000, varray);  // function to initialize the array

#pragma omp parallel private(varray)
{
    // body of parallel region not shown
}
```

Figure 6.13: **Static arrays in data environment clauses** – The compiler creates a private array with 1000 values of type `int` on the stack for each thread.


**dynamic arrays and pointers**: When working with pointers and dynamic arrays, the situation is more complicated. For example, consider the code in Figure 6.14 where we have a pointer in a `firstprivate` clause. In this case, each thread has its own private copy of the pointer, but each thread is pointing to the same physical block of storage.

```
int vptr;
vptr = (int*) malloc(1000 * sizeof(int));
initv(1000, vptr); // function to initialize the array

#pragma omp parallel firstprivate(vptr)
{
    // body of parallel region not shown
}
```

Figure 6.14: **dynamic arrays and pointers in data environment clauses** – The compiler gives each thread its own pointer pointing to the same block of memory.


You want to create a new array for each thread, but the compiler just has a pointer. It has no way of knowing which values you are interested in from the block of memory the pointer references. One solution is to use an array section. You define an array section in terms of the `lower-bound`, the `length` of the section, and the `stride`.

```
[lower-bound:length:stride]
[lower-bound:length]          // stride implied as one
[:length:stride ]             // lower-bound implied as zero
```

Using an array section in the previous example, we can have each thread allocate and copy an original variable that is an array into a parallel region with the directive:

```
#pragma omp parallel firstprivate(vptr[0:1000:1])
```

Array sections also work for the other clauses that create private copies of variables such as `private` and `reduction`.

## 6.5   Closing Comments

Understanding an OpenMP program requires that you can deduce if a variable is shared or private inside a region (such as parallel and worksharing-loop regions). Managing how variables move across region boundaries is a foundational skill for writing correct OpenMP programs.

In this chapter, we discussed the default rules for the storage attributes of variables inside a parallel region. The simplest way to remember the general rule is if a variable is in the heap memory of the process that "owns" the threads, it is shared; and if a variable is in the stack memory of each thread, it is private. And mostly (with some exceptions) if a variable is declared outside the parallel region, it is shared; and if a variable is declared inside the parallel region, it is private.

We also discussed how to modify the rules for how the storage attributes of variables change as you move between regions. To help understand these issues, we discussed a few examples. These examples demonstrated how `shared`, `private`, and `firstprivate` are used in practice.

Finally, we touched on the complex topic of debugging multithreaded programs. Of course, a parallel debugger is a great help when finding bugs in an OpenMP program. In many cases, though, you can get by without a debugger by using the `default(none)` clause. Forcing the compiler to flag each variable that moves between regions and forcing the explicit definition of its storage attribute goes a long way towards preventing sources of errors in OpenMP programs. Frankly, using `default(none)` should be standard practice when writing OpenMP programs, not a clause you fall back on once you suspect that you have bugs in your program.

# 7 Tasks in OpenMP

In Chapter 5 we discussed loop-level parallelism. To most OpenMP programmers, loop-level parallelism is the essence of OpenMP. You find the compute intensive loops in your serial application and add parallel worksharing-loop constructs to transform it into a parallel application. This style of parallel programming works best with regular problems; that is, problems where the work and the data-access patterns map directly onto a set of "potentially nested" loop indices.

The techniques from Chapter 4, fork-join parallelism with the SPMD pattern, support a wider range of algorithms. However, they still tend to favor regular problems where the challenge of defining an effective load-balancing strategy is much easier.

There is a class of important problems, however, that are irregular. They do not map directly onto nested loop indices, or if they do, the work per loop-iteration is so varied that the load balancing challenges are overwhelming. Moving through the problem domain may require following a sequence of pointers through a list, or the algorithm may be fundamentally recursive.

OpenMP in its original form ignored irregular problems. It is not that we were unaware of them, it is just that we had so much work to do to support regular parallelism that we opted to focus on those somewhat simpler problems before moving to irregular parallelism. It took us until OpenMP 3.0 to add the construct needed for irregular parallelism: the *task* construct.

In this chapter, we will explore the use of tasks in OpenMP. We will discuss the motivation behind tasks and how OpenMP had to change to support them. In addition to describing the constructs and supporting directives for task-level parallelism, we will also discuss the fundamental design patterns used with tasks.

## 7.1 The Need for Tasks

Irregular problems are numerous. They include problems based on sparse data structures with work that is highly variable, or problems with unpredictable control flows and dependencies so complex they cannot be represented by a regular iteration space. For problems based on regular data structures (such as dense arrays), if the logic is built around while loops or recursive algorithms, it can be difficult to map the algorithm onto worksharing-loops or fork-join parallelism.

Clearly there was a great need for OpenMP to support irregular parallelism. We did this by adding *tasks*, a schedulable unit of work defined by a region of code plus a data environment. They were introduced into OpenMP version 3.0 in May 2008.

Tasks were a radical enhancement to OpenMP. Prior to tasks, the central entity around which we organized the language was a thread. With tasks, we now had units of work defined independently of threads. It may sound simple, but working out the details for how tasks would mesh with a thread-centric system was very complicated. It took us over five years of dedicated work and required us to rewrite large portions of the OpenMP specification.

To understand tasks and why they are so important, let's start with a simple problem that practically demands them. Consider processing tied to the traversal of a linked list. Figure 7.1 shows the sequential code for traversing a linked list. It starts from the node at the head of the list, `head`, and walks through the list following a chain of pointer references, `p->next`. For each node, a function `processwork(p)` is called.

```
1  p = head;
2  while (p != NULL) {
3      processwork (p);
4      p = p->next;
5  }
```

Figure 7.1: **Serial linked list program**– Traverse the linked list and do a block of work (`processwork(p)`) for each node in the list where we assume `processwork(p)` for any node is independent of the other nodes.

We urge you to put the book down[1], and think about how you would parallelize this code. You cannot use a worksharing-loop construct since that only works for a `for` loop with loop-increments and loop-bounds that are invariant. There is no closed-form expression for the iteration space, meaning the length of the `while` loop is unknown at compile time and cannot be transformed into a `for` loop. It is not obvious how you would parallelize this with OpenMP since the elements of the list are data dependent and dynamic.

---

[1]A version of a linked list program is available on the book's web site, `http://www.ompcore.com`

```
1   #include <omp.h>
2   struct node {
3       int data;
4       int procResult;
5       struct node* next;
6   };
7   // initialize the list (not shown)
8   struct node* initList(stuct node* p);
9
10  // a long computation (not shown)
11  int work(int data);
12
13  void procWork (struct node* p) {
14      int n = p->data;
15      p->procResult = work(n);
16  }
17
18  int main() {
19      struct node *p = NULL;
20      struct node *temp = NULL;
21      struct node *head = NULL;
22      struct node *parr;
23
24      p = initList(p);
25
26      // save head of the list
27      head = p;
28
29      int count = 0;
30      while (p != NULL) {
31          p = p->next;
32          count++;
33      }
34      parr = (struct node*)malloc(count*sizeof(struct node));
35      p = head;
36
37      for (i = 0; i < count; i++) {
38          parr[i] = p;
39          p = p->next;
40      }
41
42      #pragma omp parallel for schedule(static,1)
43      for (i = 0; i < count; i++)
44          processwork(parr[i]);
45  }
```

Figure 7.2: **Parallel linked list program without using tasks** – Three passes through the data to count the length of the list, collect values into an array, and process the array in parallel. This is an example of the inspector-executor design pattern.

In Figure 7.2 we show one solution for parallelizing the linked list using elements from OpenMP 2.5 or earlier (i.e., before tasks were introduced to OpenMP). We carry out the list traversal in three separate steps:

- Step 1: Traverse the linked list to count the number of items in the list. Allocate an array large enough to hold the pointers to the nodes in the list.

- Step 2: Copy the pointer to each node into the array.

- Step 3: Process the nodes in parallel with a worksharing-loop construct.

This algorithm for traversing linked lists adds a great deal of overhead due to the need to traverse the data three times. The processing carried out for each node (not shown in the figures) took much longer than copying the data, so we actually observed a modest speedup with our parallel version of the program (as shown in Table 7.1). We tried the default schedule (one block of iterations per thread) and `schedule(static,1)`. The small chunk size scattered the processing for each node across the team of threads and produced a much more balanced load.

Table 7.1: **Run times in seconds for the parallel linked list program** – We run the program on a dual-core laptop with an Intel® compiler (icc) with default optimization level (O2) on Apple OS X 10.7.3 with a dual-core (four HW threads) Intel® Core™ i5 processor at 1.7 GHz and 4 GByte DDR3 memory at 1.333 GHz. The *static,1* schedule achieves a more balanced load and runs faster than the default schedule.

| Threads | Default Schedule | static,1 |
|---------|------------------|----------|
| 1       | 48               | 45       |
| 2       | 39               | 28       |

Performance, however, is not the primary concern in this case. Our simple and elegant while loop from Figure 7.2 turned into three loops and an extra array to hold pointers to each node in the list. With so many code modifications and three passes through the data structure, including copying the pointer to each node into an array, this is really cumbersome.

There has to be a better way to handle these sorts of problems; and there is. It is based on *tasks*. We will present our task-based solution to the list traversal problem later in this chapter, after we describe the basic concepts behind the task construct.

## 7.2   Explicit Tasks

The `task` construct creates an explicit task. Tasks are independent units of work composed of two parts:

- The code within the structured block associated with the task directive plus code inside any functions called within that structured block (this is called a *task region*).

- The data environment associated with the task.

The basic syntax for the task construct is shown in Table 7.2.

Table 7.2: **A task construct in C/C++ and Fortran** – The task construct creates an explicit task. The following clauses are supported in the OpenMP common core: *default(none)*, *private*, *firstprivate*, and *shared*.

| |
|---|
| **#pragma omp task** *[clause[ [, ]clause] ...]*<br>    structured block |
| **!$omp task** *[clause[ [, ]clause] ...]*<br>    structured block<br>**!$omp end task** *[nowait]* |

When a thread encounters a task construct, it may execute the task immediately or defer it for later execution. It is that *deferred* execution that makes tasks so interesting. Consider Figure 7.3. This is an illustration of how multiple tasks might execute in serial or in parallel. There are a total of 3 tasks: A, B, and C. In the serial program, they are executed one after the other. With OpenMP task parallelism, they will be placed into a task queue, and multiple threads will work through the set of tasks in parallel.

## 7.3   Our First Example: Schrödinger's Program

We wanted to start with a simple example that demonstrates how tasks work. We do this with a program we call "Schrödinger's program" as shown in Figure 7.4. The idea is to create tasks, make them wait a random amount of time, and then set a shared variable. Whichever task executes last, populates the shared variable with a value which we use to determine if Schrödinger's cat is dead or alive.

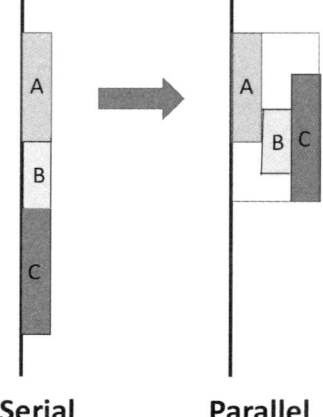

**Serial          Parallel**

Figure 7.3: **Execution of tasks in serial and in parallel** – When executing in serial, tasks are executed one after another. When run in parallel, a task may be executed at the same time as another task. The overall completion time is decreased with multiple tasks running in parallel.

The parallel region has up to two threads each of which creates two tasks. The task construct contains the code in the block following the task directive. The task region is that block plus any code called inside the function `waitAbit()`. A task either runs immediately or is deferred. When it runs, the task will wait a random amount of time and then assign a boolean variable to a shared variable `dead_or_alive`. Based on the value found in `dead_or_alive` after the parallel region has completed, the program prints that the cat is "dead" or "alive".

Note that the updates to `dead_or_alive` are not protected by synchronization constructs. Multiple threads may be updating the same location in memory at the same time. This constitutes a *data race*. In this case, we intentionally wrote a program containing a data race. We did this since we want the result of the program to be different from one run to the next without our explicit control. In other words, we want the program to contain a *race condition*. A race condition occurs when the result of a program differs depending on how the threads are scheduled by the operating system. In a shared memory programming environment, a race condition typically results from a data race.

An OpenMP program containing a data race is technically invalid. However, the

```
1    #include <stdbool.h>
2    #include <omp.h>
3    // Three functions we use but do not show here:
4    // 1. Set seed for a pseudo-random sequence
5    void seedIt(long *val);
6    // 2. function to flip a coin (randomly return true or false)
7    bool flip(long *coin);
8    // 3. wait a short random amount of time
9    double waitAbit();
10
11   int main()
12   {
13       double wait_val;
14       long rand, i;
15       int dead_or_alive;
16       omp_set_num_threads(2);
17
18       // "flip a coin" to choose which task is for the dead
19       // cat and which for the living cat.
20       long coin;
21       seedIt(&coin);
22       bool HorT = flip(&coin);
23
24       printf("Schrodinger's program says the cat is");
25
26       #pragma omp parallel shared(HorT, dead_or_alive)
27       {
28           // These tasks are participating in a data race
29           #pragma omp task
30           {
31               double val = waitAbit();
32               dead_or_alive = HorT;
33           }
34           #pragma omp task
35           {
36               double val = waitAbit();
37               dead_or_alive = !HorT;
38           }
39       }
40       if (dead_or_alive)
41           printf(" alive. \n");
42       else
43           printf(" dead. \n");
44
45       return 0;
46   }
```

Figure 7.4: **Schrödinger's Program** – Two threads each generates two tasks. They wait a random bit of time and then set a shared variable to true or false. Whichever task executes last determines the final value of the variable and whether the cat is "dead" or "alive".

data race in this example code is "benign". In the C programming language, when an `int` is used as a logical variable, any value other than 0 is true. Therefore, even if the data race led to corrupted values of the shared variable, the behavior of the program would still match expectations. It randomly selects a cat as being "dead" or "alive". The data race does not prevent the program from accomplishing its job; hence why we call it a "benign" data race[2].

We would be remiss if we did not point out that the phrase "benign data race" is controversial and perhaps even an oxymoron. Technically, there is no such thing as a data race that is "benign" since modern programming languages, including OpenMP, define any program with a data race as invalid. Many of us in the parallel applications community, however, were quite unhappy when language designers decided to forbid benign data races. It is rare, but there are algorithms that benefit greatly from them[12]. For example, in some iterative solvers based on relaxation methods[8], we might omit expensive synchronization constructs needed to prevent extremely rare data races. If a data race happens to corrupt a value, subsequent iterations will clean it up. This method works and has been successfully used for many years. Hence, our inclusion of benign data races in this book is our way of keeping the fight alive. At the same time, we recognize that deliberately putting a data race into a program is not generally advised and should either NOT be done or done with extreme caution.

## 7.4   The Single Construct

Before we discuss how tasks are used in practical programs, we need to introduce an additional OpenMP construct. This is the *single* worksharing construct the syntax of which is described in Table 7.3. The *single* construct denotes a block of code that is executed by only one thread. That thread is any thread in the team, usually the first that happens to encounter the construct. As with all the worksharing constructs, the `single` construct implies a barrier at the end of the construct. The thread that encounters the single construct does the work inside the construct, while the other threads wait at the implied barrier at the end of the construct. Once the thread executing the single construct has completed its work, all threads continue their execution beyond the single construct.

---

[2]Ideally for the race to be benign, this program should run on a machine with word-level atomicity; i.e., loads and stores for single-word variables cannot be observed in an intermediate state.

Table 7.3: **A single construct in C/C++ and Fortran** – The single construct is a worksharing construct executed by one thread in a team while the other threads wait at the barrier implied at the end of the construct. This barrier can be disabled by the use of a *nowait* clause.

| |
|---|
| **#pragma omp single** *[nowait]*<br>    structured block |
| **!$omp single**<br>    structured block<br>**!$omp end single** *[nowait]* |

We provide an example of how the `single` construct is used in Figure 7.5. This is a common pattern when working with a hybrid MPI/OpenMP program. A team of threads cooperates to do some work (the function *do_many_things()*). Since some implementations of MPI do not work well with multiple threads, we designate one thread to work with MPI to exchange boundaries between MPI ranks (with the MPI calls inside the function *exchange_boundaries()*). The other threads wait at the barrier implied at the "curly brace" at the end of the `single` construct, then when the boundary exchange is complete, all the threads continue in parallel to call *do_many_other_things()*.

```
1  #pragma omp parallel
2  {
3      do_many_things();
4      #pragma omp single
5          {
6              exchange_boundaries();
7          }
8      do_many_other_things();
9  }
```

Figure 7.5: **An OpenMP single construct example** – All threads execute *do_many_things* and *do_many_other_things*, but only one thread executes *exchange_boundaries*.

As with the worksharing loop construct, you can disable the implied barrier at the end of the construct by using a `nowait` clause. Be extremely careful when using a `nowait`. It is remarkably easy to introduce a data race when you disable the barrier

at the end of a worksharing construct. These sorts of errors are easy to introduce
and extremely difficult to track down.

## 7.5  Working with Tasks

The explicit task construct is very flexible and can be used in many ways. By
far the most common pattern when working with the task construct, however, is
to have one thread create the tasks while the other threads wait at a barrier and
execute the tasks. We show an example of this pattern in Figure 7.6. One thread is
selected to create the tasks through a single construct. That thread defines explicit
tasks `fred()`, `daisy()`, and `billy()` while the other threads in the team wait at
the barrier at the end of the single construct. As tasks are deferred for execution
(i.e., placed in a task queue), the tasks waiting at the barrier compute the tasks.
When the thread executing the region in the single construct and all the tasks have
completed, the team of threads continues past the end of the single construct.

```
1   #pragma omp parallel
2   {
3       #pragma omp single
4       {
5           #pragma omp task
6               fred ();
7           #pragma omp task
8               daisy ();
9           #pragma omp task
10              billy ();
11      } //end of single region
12  } //end of parallel region
```

Figure 7.6: **A basic task example** – Inside a parallel region, 3 tasks are created by a
single thread.

While this pattern is by far the most common way you will see tasks used, it is
perfectly OK to have multiple threads create tasks in parallel. We have already seen
an example of threads creating tasks in parallel with our Schrödinger's program in
Figure 7.4.

An OpenMP implementation is expected to keep track of how many tasks have
been placed in the task queue. Rather than letting a task queue overflow and
leading to potentially catastrophic failure of the program, the OpenMP runtime will

suspend the thread creating tasks and then use it to help the other threads work on other tasks. This would continue until the task queue has been drained at which point the thread would switch back to its work creating new tasks.

Tasks are ideal for irregular problems. A task-oriented system is said to automatically balance the load (an example of automatic, dynamic load balancing). Threads waiting at a barrier pull a task from the task queue, carry out the work, and then go back to the queue for another task. This automatically maintains an even distribution of work across the threads as long as there are many more tasks than threads (which is the normal situation).

### 7.5.1   When Do Tasks Complete?

You create explicit tasks to do the work defined by an algorithm. As you move through different phases of an algorithm, you need to know when you can assume tasks are complete. To describe this in more detail, we need to define some key terms with tasks.

A set of tasks created directly within the same construct are called *sibling* tasks. They are created by a single parent task, so it follows naturally to call them siblings; that is, they are all *child* tasks of the same parent. For example in Figure 7.6, the tasks `fred()`, `daisy()`, and `billy()` are sibling tasks. It is easy to imagine that inside the tasks `fred()`, `daisy()`, and `billy()` they each create additional tasks. In other words, it is quite common to have nested task constructs, where one task construct creates additional tasks. Consider a task, we will call it *taskA*, that creates other tasks. These are the child tasks of taskA. These tasks may create other tasks which may go on to create even more tasks. We call all of those tasks *descendant* tasks of taskA.

We now can describe when tasks complete. All tasks (sibling tasks and their descendent tasks) complete before any threads move beyond a barrier. This barrier is often the barrier at the end of a `single` construct. However, if there is a *nowait* clause on the single clause, the tasks will all complete at or before the next barrier, which is typically the *parallel* construct at the end of a parallel region. Therefore, all the tasks generated inside a parallel region will complete before the parallel region ends.

There are times you want more fine-grained control over the execution of a thread relative to the completion of its tasks. The *taskwait* directive indicates that the thread encountering the taskwait, will pause until its child tasks have completed;

that is, the thread will wait at the *taskwait* until the tasks *it* created prior to the *taskwait* have completed. Figure 7.7 shows an example of the *taskwait* directive.

```
1   #pragma omp parallel
2   {
3      #pragma omp single
4      {
5         #pragma omp task
6            fred ();
7         #pragma omp task
8            daisy ();
9         #pragma omp taskwait
10        #pragma omp task
11           billy ();
12     }
13  }
```

Figure 7.7: **A taskwait example** – Tasks *fred* and *daisy* must complete before task *billy* starts.

This example is very similar to the example shown in Figure 7.6 except now after the tasks for `fred` and `daisy` are generated, there is a taskwait. This causes the thread that encountered the taskwait directive to wait until `fred` and `daisy` have completed their work and have returned. Then that thread can continue and generate the task `billy`.

To reiterate, threads waiting at a barrier will proceed once all outstanding tasks have completed. This includes sibling tasks and children of those sibling tasks. If you want to wait on just sibling tasks created up to a point, that is, those tasks all created in the same lexical extent up to a fixed point, you use `taskwait`.

## 7.6   Task Data Environment

### 7.6.1   Default Data Scoping for Tasks

Data scoping rules for tasks are similar to the rules for other OpenMP constructs discussed in Chapter 6. There are two major differences:

1. The data environment binds to the task, not the thread encountering the task. This is important since you do not control which thread executes which task.

2. If a variable is *private* when encountering the task, it will be made *firstprivate* by default.

Both of these rules are essential since a task can be deferred. You need to capture the value of variables when the task is generated. These variables could even be "out of scope" by the time a deferred task executes, so we need to make sure that the original data environment is captured when a task is created.

Hence, we can summarize the key rules for the default data environment: Variables that are *private* when the task construct is encountered are *firstprivate* by default; Variables that are *shared* in all constructs starting from the innermost enclosing parallel construct are *shared* by default.

Let's use an example to elaborate on these rules. In Figure 7.8, A is shared and B is private when the parallel region is created. Then inside the task construct, A is still shared, because a shared attribute remains intact as you move between task regions. It is shared inside and outside of the task.

```
1   #pragma omp parallel shared(A) private(B)
2   {
3       ...
4       #pragma omp task
5       {
6           int C;
7           compute(A, B, C);
8       }
9   }
```

Figure 7.8: **Tasks data environment example** – *A* is shared, *B* is firstprivate, and *C* is private.

C is declared inside the task construct, it is *private*. This is just the familiar rule from other OpenMP constructs. If a variable is declared inside a block of code, it is private. After we exit from the block of code, the private variable will be out-of-scope and unavailable.

B is private when encountering the task region, it is *firstprivate*. Since a task can be deferred and much can happen to the value of a variable between the time a task is generated and when it executes, it will be made firstprivate by default since that is the safe behavior. Firstprivate will create a private variable and it will initialize it from the original variable at the time the task construct was encountered. This feature is extensively used with OpenMP tasks.

In summary, regarding data scoping for tasks, variables can be shared, private or firstprivate with respect to task, just as we have for any other OpenMP constructs. The concepts below are a little bit different compared with threads:

- The data environment binds to the task, not the thread encountering the task.

- If a variable is shared on a task construct, the references to it inside the construct are to the same address space of this shared variable when the task was encountered.

- If a variable appears in a private clause on a task construct, the references to it inside the construct are to a new uninitialized storage that is created when the task is executed.

- If a variable is firstprivate on a task construct, the references to it inside the construct are to a new storage with the same variable name that is created and initialized with the value of that variable when the task is encountered.

### 7.6.2   Linked List Program Revisited with Tasks

Now we are ready to revisit the linked list program and parallelize it using OpenMP tasks. Figure 7.9 shows this elegant and simple solution, compared to the cumbersome three-pass solution in Figure 7.2.

The OpenMP parallel construct creates a team of threads. A single thread packages the tasks, and the other threads wait at the barrier, and work on the tasks being put on the queue.

A single thread first grabs the head of the list, then walks through the linked list with the while loop, and packages the process of each node into a task, then goes on to the next node until the list is exhausted. Once the thread that executes the single region has completed creating the tasks, it will join the other threads to execute the tasks.

Notice the `firstprivate` clause is used for the `task` directive in line 8. This is because we need to capture the relevant data environment at the time when the task is created, and package that up with the tasks.

```
1   #pragma omp parallel
2   {
3      #pragma omp single
4      {
5         p = listhead ;
6         while (p)
7         {
8            #pragma omp task firstprivate(p)
9            {
10              process (p);
11           } // end of task creation
12           p = p->next;
13        }
14     } // end of single region
15  } // end of parallel region
```

Figure 7.9: **Linked list with tasks** – The implementation with OpenMP tasks is much more elegant than the three-pass solution in Figure 7.2.

## 7.7   Fundamental Design Patterns with Tasks

We often use a "Hello World" program as our first program, be it for a serial C or Fortran code, for a parallel MPI or OpenMP code, or for any other programming languages. The analogous "Hello World" program for parallel programming with regular problems is our Pi program. For the irregular applications, the Fibonacci program in Figure 7.10 is our "Hello World" program. We know this is a terrible way to compute Fibonacci numbers. Nonetheless, it is an excellent way to demonstrate the key design patterns for working with tasks so parallel computing educators use it often.

Fibonacci series are defined as `F(n)` = `F(n-1)` + `F(n-2)` with given initial values for when `n` = 0 and `n` = 1. Although this code shows an inefficient $O(n^2)$ recursive implementation, it is suitable for illustrating OpenMP tasks concepts for irregular applications.

In this code, it first defines the base case which terminates the recursion. In this case, it holds when `n` < 2. Then for larger `n` values, it calls `x=fib(n-1)`, `y=fib(n-2)`, and returns `x + y`. Here the `fib(n-1)` will recursively call `fib(n-2)` and `fib(n-3)`, until it reaches the base case.

Let's consider how to parallelize the Fibonacci program using OpenMP tasks. Figure 7.11 shows the parallel implementation. Each computation of `fib(n-1)` and

```
1    int fib (int n)
2    {
3        int x,y;
4        if (n < 2) return n;
5
6        x = fib(n-1);
7        y = fib(n-2);
8        return (x+y);
9    }
10
11   int main()
12   {
13       int NW = 5000;
14       fib(NW);
15   }
```

Figure 7.10: **Fibonacci example** – This is the serial recursive implementation.

fib(n-2) can be treated as a task, and each task can create subtasks to compute fib(n-2) and fib(n-3), etc.

We recursively generate function invocations and build a binary tree of tasks. A task cannot complete until all tasks below it in the tree are complete, which is enforced with *taskwait*. The *taskwait* makes sure that both the *fib(n-1)* and *fib(n-2)* tasks are completed, before they can be combined to return the result for *fib(n)*.

In order for x and y to be available outside the data environment of each task they must be shared. They are declared inside the `fib()` function (on line 3) which makes them private to the thread that encounters the task constructs on lines 6 and 7. Hence we must use **shared** clauses on the task directive to force them to be shared.

Notice in the main program, `fib(NW)` is called by a single thread within a parallel region. The single thread creates the tasks to compute x and y, and the other threads wait at the barrier and work on the tasks and subtasks that are created. However, if `fib(NW)` was not called inside a parallel region, there would be no threads to execute the program in parallel. We say that the **single** construct is an orphaned construct, that is, an OpenMP construct without a team of threads to execute it.

```
1   int fib (int n)
2   {
3       int x,y;
4       if (n < 2) return n;
5
6   #pragma omp task shared(x)
7       x = fib(n−1);
8   #pragma omp task shared(y)
9       y = fib(n−2);
10  #pragma omp taskwait
11      return (x+y);
12  }
13
14  int main()
15  {
16      int NW = 5000;
17      #pragma omp parallel
18      {
19          #pragma omp single
20              fib(NW);
21      }
22  }
```

Figure 7.11: **Parallel implementation of the Fibonacci program using OpenMP tasks** – Two tasks create child tasks recursively. `taskwait` ensures the direct child tasks complete before the merge. The base case to exit the recursion is defined for when `n < 2`.

### 7.7.1   Divide and Conquer Pattern

The algorithm used in the Fibonacci program is the classic divide and conquer pattern. This is an essential pattern for recursive programs, just as the loop-level parallelism pattern is fundamental to grid-based, regular mesh codes.

With the divide and conquer pattern, we recursively split the problem into smaller and smaller sub-problems, continuing until the sub-problems become so small that it makes sense to just solve them directly. Then we reverse the process taking our directly solved sub-solutions and merging them up the tree to generate the final solution. Figure 7.12 illustrates this divide and conquer pattern.

One of the decisions to make when implementing a divide and conquer algorithm is when to do the direct solve. In our Fibonacci program, we split all the way down to `fib(1)` and `fib(2)`. Usually, you define the base case (i.e., the size problem where you solve directly) to be much larger, to balance the cost of splitting into

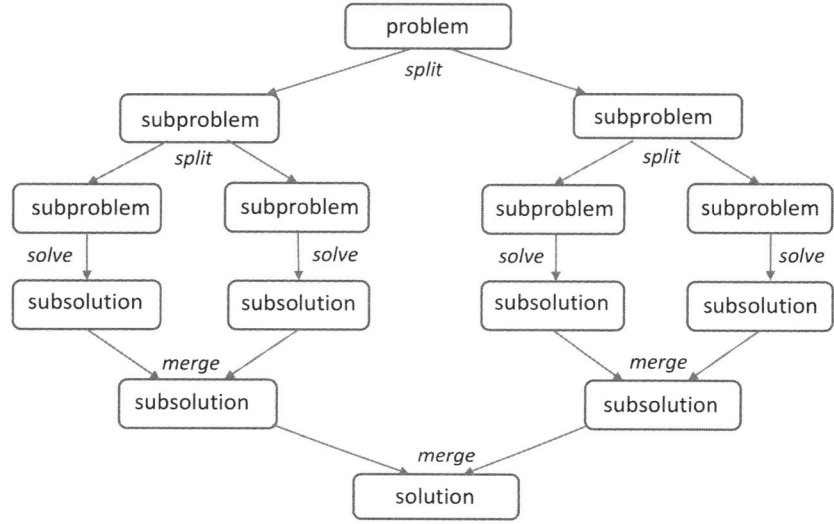

Figure 7.12: **Illustration of the divide and conquer pattern** – The pattern breaks down into three parts: a recursive splitting into a tree of subproblems, a direct solve, and then merging "back up the tree" to obtain the overall solution.

smaller and smaller subproblems compared to the cost of a direct solve. For example, in a divide and conquer linear equation solver, you might choose the base case to be the size of the problem that just fills the last level cache of your processor.

In general with this pattern, there are 3 options for doing work: you can do it as you split into sub-problems, or you can do work only at the leaves (at the solve step), or you can do work as you recombine (as you merge sub-solutions). Parallelism emerges naturally from this pattern as each of the individual solve, split, and merge steps can usually proceed in parallel.

The Fibonacci problem is one of the more straightforward examples of the divide and conquer pattern. Other examples including dynamic programming, symbolic reasoning programs, and even finite element codes.

We will use our numerical integration program (i.e., our Pi example) to show how the divide and conquer pattern can be used with loop oriented algorithms. We present the now familiar loop-based Pi program again in Figure 7.13.

We need to think about how we will split the full problem into sub-problems, what will the base case be (when we stop splitting into smaller sub-problems) and how the merge step will work.

```c
#include <stdio.h>
#include <omp.h>
static long num_steps = 1024*1024*1024;
double step;
int main()
{
    int i;
    double x, pi, sum = 0.0;
    double start_time, run_time;

    step = 1.0 / (double) num_steps;

    start_time = omp_get_wtime();

    for (i = 0; i < num_steps; i++) {
        x = (i + 0.5) * step;
        sum += 4.0 / (1.0 + x * x);
    }

    pi = step * sum;
    run_time = omp_get_wtime() - start_time;
    printf("pi = %lf, \%ld steps \%lf, \%lf secs\n ",pi,
                    num_steps, run_time);
}
```

Figure 7.13: **Serial Pi program to numerically estimate a definite integral using the midpoint rule** – The loop iterations are independent other than the summation into sum.

We can split this problem into subproblems by splitting the loop in half. To avoid odd boundary conditions in splitting the loop, we have defined the number of steps in the integration to be a large power of 2. We define a function that computes a partial sum over a contiguous subset of the loop iterations:

```c
double pi_comp(int Nstart, int Nfinish, double step)
```

Inside this function, we implement our recursive split:

```c
long iblk = Nfinish - Nstart;
sum1 = pi_comp(Nstart, Nfinish - iblk/2, step);
sum2 = pi_comp(Nfinish - iblk/2, Nfinish, step);
```

Now we need to decide the base case. This is a parameter you vary when optimizing your final program. We know from experience that the overhead of the split phase is large compared to the computation, so we make our base case quite large. We represent the base case with a parameter MIN_BLK and test if the size of a subproblem is less than MIN_BLK in which case we would solve directly. We put everything together into a single serial program in Figure 7.14.

```
1   #include <omp.h>
2   static long num_steps = 1024*1024*1024;
3   #define MIN_BLK 1024*256
4   double pi_comp(int Nstart, int Nfinish, double step)
5   {
6       int i, iblk;
7       double x, sum = 0.0, sum1, sum2;
8       if (Nfinish - Nstart < MIN_BLK){
9           for (i = Nstart; i < Nfinish; i++) {
10              x = (i + 0.5) * step;
11              sum += 4.0 / (1.0 + x * x);
12          }
13      }
14      else {
15          iblk = Nfinish - Nstart;
16          sum1 = pi_comp(Nstart, Nfinish - iblk/2, step);
17          sum2 = pi_comp(Nfinish - iblk/2, Nfinish, step);
18          sum = sum1 + sum2;
19      }
20  return sum;
21  }
22
23  int main ()
24  {
25      int i;
26      double step, pi, sum;
27      step = 1.0 / (double) num_steps;
28      sum = pi_comp(0, num_steps, step);
29      pi = step * sum;
30  }
```

Figure 7.14: **Serial Pi program using the divide and conquer pattern** –Just to make the code simpler, we pick a number of steps that is a power of 2. This way we can split the number of steps in half repeatedly and always create intervals that are divisible by 2.

Start by looking at the split step inside the pi_comp() function. You need to understand how the subproblems are formed from the input terms that define the problem. Then implement the base case. As you can see, this is just the familiar loop from the Pi program which we use when the number of iterations in a subproblem falls below MIN_BLK. Then as a last step we write the main program which launches the recursive process.

At this point, we urge you to stop reading. Go to Figure 7.14 and see if you can see where to add OpenMP directives to create a task parallel version of this program[3]. Look back at the Fibonacci program in Figure 7.11 if you need some hints to help you solve this problem. The solution is shown in Figure 7.15.

The overall structure follows directly from the serial code. You have to create a team of threads in the main program. Then one thread inside the team (using a single construct) starts the recursive process. The split phase creates two tasks to calculate sum1 and sum2. Both of these need to be shared variables, so that they will still exist after the tasks are done. Taskwait ensures these tasks are completed before sum1 and sum2 are merged.

Think carefully how this recursive task-based code will execute. Because of the taskwait directives after each split, you build a tree of tasks blocked and waiting for sub-solutions in order to proceed. Picture a tree of tasks waiting for sub-solutions to drive the merge. Finally, the recursive splitting reaches the base case where the sub-problem is directly solved. At this point, the tree collapses as sub-solutions release waiting tasks and the problem quickly merges into the full solution.

Table 7.4 shows Pi results using tasks. The performance from Pi tasks is surprisingly good compared with the other parallel algorithms we have examined in previous chapters.

Table 7.4: **Run times in seconds for the numerical integration program with and without array padding, using a critical section, using parallel worksharing-loop, and using parallel tasks** – Serial program ran in 1.83 seconds.

| Threads | 1st SPMD | 1st SPMD Padded | SPMD Critical | Pi Loop | Pi Tasks |
|---------|----------|-----------------|---------------|---------|----------|
| 1 | 1.86 | 1.86 | 1.87 | 1.91 | 1.87 |
| 2 | 1.03 | 1.01 | 1.00 | 1.02 | 1.00 |
| 3 | 1.08 | 0.69 | 0.68 | 0.80 | 0.76 |
| 4 | 0.97 | 0.53 | 0.53 | 0.68 | 0.52 |

---

[3]The recursive Pi program is available on our Common Core web site, http://www.ompcore.com

```
1   #include <omp.h>
2   static long num_steps =   1024*1024*1024;
3   #define MIN_BLK 1024*256
4   double pi_comp(int Nstart, int Nfinish, double step)
5   {
6       int i, iblk;
7       double x, sum = 0.0, sum1, sum2;
8       if (Nfinish − Nstart < MIN_BLK){
9           for (i = Nstart; i < Nfinish; i++){
10              x = (i + 0.5) * step;
11              sum += 4.0 / (1.0 + x*x);
12          }
13      }
14      else {
15          iblk = Nfinish − Nstart;
16          #pragma omp task shared(sum1)
17              sum1 = pi_comp(Nstart, Nfinish − iblk/2, step);
18          #pragma omp task shared(sum2)
19              sum2 = pi_comp(Nfinish − iblk/2, Nfinish, step);
20          #pragma omp taskwait
21              sum = sum1 + sum2;
22      }
23  return sum;
24  }
25
26  int main()
27  {
28      int i;
29      double step, pi, sum;
30      step = 1.0 / (double) num_steps;
31      #pragma omp parallel
32      {
33          #pragma omp single
34          sum = pi_comp(0, num_steps, step);
35      }
36      pi = step * sum;
37  }
```

Figure 7.15: **Parallel Pi program using tasks** – It is accomplished with the divide and conquer pattern by splitting the problem into two subtasks to calculate *sum1* and *sum2*, recursively solving each task, and then combining the results.

In this particular case, the performance with tasks is on par with the best SPMD results. It even beats the results with loop-level parallelism. This is an unusual case. Usually, you should not use tasks for situations already well supported in OpenMP. Tasks incur considerable extra overhead due to the single thread that generates the tasks. The runtime system must assign tasks to threads to manage the task queue. And finally, the task runtime system has to support the synchronization process to wake up tasks waiting for other tasks to complete.

## 7.8   Closing Comments

In this chapter, we talked about regular *vs.* irregular problems. Originally OpenMP was focused on regular problems that could be described in terms of loop-nests. There are many problems, however, that are irregular; that do not fit onto a basic index space defined by a set of loops. OpenMP needed to grow to support these problems. We needed tasks.

We then described tasks; what they are. In particular, a task is code and a data environment. This data environment shares a great deal with the data environments we have learned about for threads, but since a task can be deferred, the rules about how variables move across data environments are slightly different. The biggest difference is that for original variables that are private, OpenMP makes them firstprivate for a task. In other words, it captures the values of those variables at the time the task is generated, so the values are there when needed for the computation.

We then explained the divide and conquer pattern for a recursive Fibonacci program. This simple program covers most of the concepts needed to apply the divide and conquer pattern to real problems. To drive that point home, we went through the process of applying the divide and conquer pattern to a second problem: our Pi program used extensively earlier in the book.

Tasks are great. They greatly expand the range of problems that can be addressed with OpenMP. Task management, however, adds overhead. Our recommendation is do not use tasks if there is a natural way to solve your problem using the rest of the OpenMP Common Core. Do not expect miracles from the runtime system. When working with tasks, expect to spend time optimizing for the number and granularity of tasks.

# 8 OpenMP Memory Model

We use OpenMP to write multithreaded programs. Threads share memory and execute concurrently; that is the instructions from different threads are not ordered with respect to each other. This means the threads can deal with many things at once since "unordered" means they can be active and make progress at the same time. With multiple processors for those threads to run on, those concurrent threads will run in parallel and the program's work will complete in less time.

Concurrency is fundamental to OpenMP. When operations in concurrent threads only read values from the shared memory and write address ranges distinct to each thread (i.e., they do not overlap), writing a multithreaded program that is both faster and correct is straightforward. When the concurrent threads are both reading and writing to overlapping addresses in shared memory, however, a programmer must define an order to the shared memory operations arising from different threads so they do not conflict.

If two or more threads execute a mix of reads and writes to the same addresses in memory and those read and write operations are not ordered by a synchronization operation, the program has a *data race*. The final value in memory is undefined for addresses involved in a data race. That means the result of the program is undefined and the program is invalid.

One solution is to only write multithreaded programs that do not mix reads and writes to overlapping regions of shared memory. These are called *embarrassingly parallel programs*. They are easy to write and can deliver impressive speedups. Unfortunately, most algorithms are not embarrassingly parallel. Most problems are organized around data structures that need to be updated in parallel. That translates into multiple concurrent threads reading and writing to overlapping ranges of addresses in memory.

We have alluded to this problem many times. In discussing synchronization we described `barriers` and other OpenMP constructs (such as `critical`) that manage memory conflicts and support the writing of race-free programs. In this chapter, we revisit this problem and provide some of the lower-level details needed to understand how to write correct programs that mix reads and writes to overlapping addresses in shared memory by concurrent threads. Consistent with the guiding principles behind the OpenMP Common Core, we will focus on a restricted subset of the rules needed to work with shared memory. They are sufficient to the needs of most parallel application programmers, but avoid some of the more complicated issues advanced concurrent algorithm designers may need. Those more advanced shared memory topics are left to Chapter 11, when we move "beyond the Common Core".

## 8.1   Memory Hierarchies Revisited

We introduced the memory hierarchy of a typical multiprocessor CPU in Section 1.3.1. In this section we will revisit the multiprocessor CPU and consider how it interacts with the values of variables when viewed from multiple concurrent threads. To simplify the discussion, we will consider a dual-core CPU. We show a schematic diagram of a dual-core CPU in Figure 8.1.

A *variable* is a name for an address in memory. This memory may be virtual memory or physical memory. For now, let's specialize the discussion to physical memory. That means a variable is a location in Random Access Memory (RAM). In most systems RAM is implemented with Dynamic Random Access Memory (DRAM). As we discussed in Section 1.3.1, the time to access a value in DRAM is long relative to the clock on a CPU, so the memory hierarchy consists of a hierarchy of faster memories close to the processor cores. The fastest memory is the register file directly accessed by the low level instruction set of the CPU.

The next levels of the hierarchy are the caches. A memory location in cache is associated with a contiguous block of variables at a location in memory, the so-called *cache line*. A cache is not a separate address space. It is a buffer that holds temporary copies of values from the shared address space (the Shared memory). There are one or more levels of cache tightly integrated and local to each core. Then there is typically a larger cache furthest from the cores (the *last level cache*) that in the case of Figure 8.1 is shared among the cores.

Consider a variable $\gamma$ in Figure 8.1. There is only one address in the memory and a single value represented by the bytes at that address. Throughout the memory hierarchy, from the register file on down through the various levels of cache, there are temporary values for the variable $\gamma$. A cache coherence protocol manages these values and assures that over time they provide a common view of memory. At any given moment, however, the values for $\gamma$ may be inconsistent. In other words, the value sitting in a register may be different than the value in the various levels of cache which may be different from the value in DRAM. The topic of *memory consistency* addresses these values and how they vary with respect to each other at a fixed point in time. When they are allowed to differ at any given point in time, we say that the system has a *relaxed memory consistency model*.

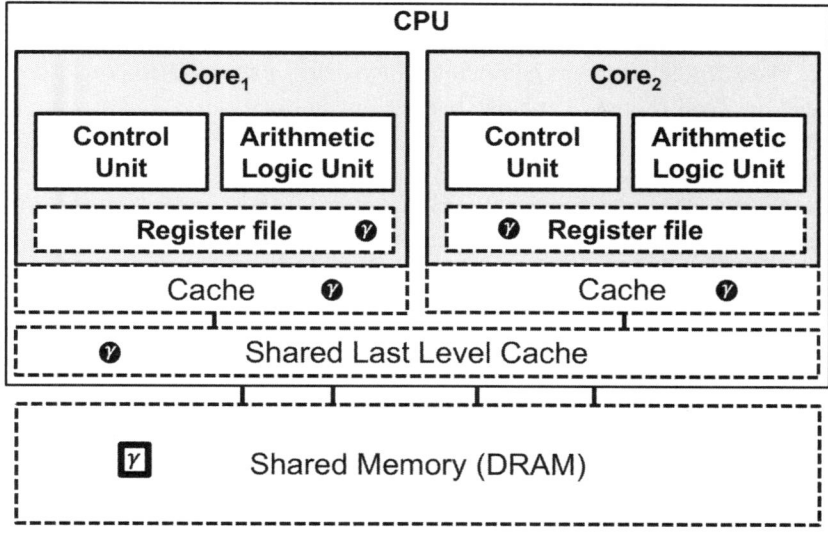

Figure 8.1: **A simplified view of a dual-core CPU with the memory hierarchy highlighted through use of dashed boxes** – A variable $\gamma$ represents a specific address in the shared memory and is shown here as a square box. At any given time, the value associated with this address may exist at each level in the memory as shown by the variable name in a black circle.

Any system you are likely to use with OpenMP uses a relaxed memory consistency model. Consider the alternative, that is, a system where the values for a given variable across the memory hierarchy were forced to be consistent at every point in time. This would require enforcing a fixed order to updates of shared variables and moving copies of a variable across the memory hierarchy each time the variable is updated. Code updating shared variables would have to be serialized which would create huge overheads for all but embarrassingly parallel programs. Therefore, multiprocessor systems use relaxed memory consistency models. Any programming model suited to multiprocessor systems must grapple with this reality.

A relaxed memory model can lead to some strange results. Consider the program in Figure 8.2. For thread 0 the program order indicates that x is set to 1 followed by an assignment of x to the variable r. You would think that on thread 1, if it saw a value of r==1 it could only see a value of 1 for x in the assignment to y. The values of the variables across the memory hierarchy, however, are not forced to be consistent between the two threads. It does not happen often, but if you run this program repeatedly, cases will arise where r == 1 and x == 0 on thread 1. The order of reads and writes to the shared variables r and x are unordered between the two threads. Without synchronization constructs to impose a fixed order, the program contains a data race and is invalid.

```
1   #include <omp.h>
2   #include <assert.h>
3
4   int main()
5   {
6       int x = 0, y = 0, r = 0;
7       omp_set_num_threads(2);     // request two threads
8       #pragma omp parallel
9       {
10          int id = omp_get_thread_num();
11          #pragma omp single
12          {
13              int nthrds = omp_get_num_threads();
14
15              // verify that we have at least two threads
16              if (nthrds < 2) exit(1);
17          } // end of single region
18
19          if (id == 0) {
20              x = 1;
21              r = x;
22          }
23          else if (id == 1) {
24              if (r == 1) {
25                  y = x;
26                  assert(y == 1);   // Assertion will occasionally fail;
27                                    // i.e., r == 1 while x == 0
28              }
29          }
30      } // end of parallel region
31  }
```

Figure 8.2: **A program with a race condition** – A relaxed memory model permits the assertion to fail; i.e., the thread with id == 1 can observe values in memory such that r ==1 while x is still 0.

The set of rules used by a programming language that defines the value that can be loaded when a shared variable is read is called the *memory model*. The OpenMP memory model is based on the memory model defined in the C++ programming language. In the OpenMP version 5.0 specification, it occupies many pages of dense text spread out across the specification, text that will tax the ability of most experienced parallel programmers to fully understand. For the Common Core, we define a subset of the memory model. It is easy to understand and for most application programmers, is all they ever need.

## 8.2   The OpenMP Common Core Memory Model

The original memory model for OpenMP (from version 1.0 of the specification) was defined in terms of an operation called *flush*. A flush forces a thread's temporary view of its variables to be consistent with the values of the variables in memory (i.e., in RAM). Variables that are being read are marked as invalid so the next time they are accessed, they will be loaded from memory (as opposed to a register or in cache). Variables written by a thread sitting in cache, a register file, or a write buffer of any kind are written to memory.

A flush applies to all of the variables that are shared among threads. This set of variables is called the *flush set*. There are ways in OpenMP to define a flush set that contains only a subset of the shared variables but using this form of the flush is extremely difficult to use correctly and is not advised for any but the most advanced programmers.

The next ingredient to understanding the OpenMP memory model is to consider how compilers reorder operations around a flush. The statements in a program's text define a series of loads and stores to memory. We call this the *program-order*. A compiler will reorder those loads and stores to optimize performance. This is the *compiler-order*. Modern microprocessors may at runtime further reorder the operations. This is the *execution-order*. These reorderings have a dramatic impact on performance. We would all be seriously unhappy with the performance of our systems if these reorderings were disabled.

A compiler, working closely with the microprocessor architecture, promises programmers that they will only observe the program-order for memory operations. The compiler-order and execution-order may be radically different from the program-order, but these differences must not be observable. A compiler, however, only understands the execution of a single thread when considering the sequence of instructions in a program. When multiple threads are mixing loads and stores to an overlapping set of variables, the compiler-reordered operations could break a program by introducing data races.

Hence, the memory model of OpenMP restricts how a compiler can reorder operations when using variables from the flush set. In a program that is *properly synchronized*, you cannot observe an execution-order for which any variables from the flush set are moved around a flush. This means that writes issues before a flush must complete and be written to memory before the flush operation completes. Likewise, reads and writes that follow a flush cannot occur until the flush completes.

With the flush and constraints on compiler reordering operations around a flush, threads can safely mix reads and writes to a shared variable if they carry out the following steps *in order*:

1. A value is written to a shared variable by the first thread.

2. The first thread issues an OpenMP flush operation.

3. The second thread issues an OpenMP flush operation.

4. The second thread reads the value of the shared variable.

The key word here is "in order". The operations on the first thread will always be observed to run in program order. The operations on the second thread likewise will always be observed to run in program order. An ordering constraint, that is a *synchronization operation*, is needed to assure that the flushes on *both* threads occur between the write on thread 1 and the read on thread 2. This is what we mean when we use the phrase *properly synchronized*.

A flush is a per thread operation. A thread issues a flush operation and the values for the variables in the flush set for the thread issuing the flush operation are made consistent with shared memory. One thread flushing its flush set says nothing about the values of a flush set on another thread. Hence, all threads involved in sharing values of variables must issue the flush operations.

The flush is not a synchronization operation. A synchronization operation defines an ordering constraint among two or more threads. A flush only impacts the memory operations of a single thread and says nothing at all about what any other threads are doing. The flush is an essential aspect of synchronization, but the flush itself is not a synchronization operation.

Therefore, to safely mix reads and writes to an overlapping range of values in shared memory, you need both the flush operation (to force memory consistency) and the synchronization operation. In other words, you need synchronization operations to make sure the reads and writes are correctly ordered with respect to the flushes. For the OpenMP Common Core, that means you need to use either a barrier or a critical construct to safely order the potentially conflicting reads and writes.

Advanced programmers looking to aggressively minimize parallel overheads carefully lay flushes around synchronization operations. It turns out, however, that doing so correctly is extremely difficult. Hence, while we describe the concept of a flush and use it to define the OpenMP memory model, we do not include an explicit flush in the OpenMP Common Core. Instead, we take advantage of the fact that OpenMP implies a flush where they are needed. You get the needed features from the flush without having to explicitly place the flush in various threads.

A flush is implied at the following points in the OpenMP Common Core:

- When a new team of threads is forked by a `parallel` construct.

- When a critical construct is entered by a thread.

- When a thread completes a critical section and exits the critical section.

- On entry to a task region.

- On exit from a task region.

- On exit from a taskwait.

- On exit from an explicit barrier.

- On exit from an implicit barrier:

    - At the end of a parallel region.
    - At the end of a worksharing construct (a worksharing-loop or a `single`) unless disabled with a `nowait` clause.

If programmers manage ordering constraints between reads and writes to shared variables using the above list, they can be confident their programs are race free. The OpenMP implementation takes care of the flushes for you. In practical terms, this leads to the simple memory model summarized in the boxed text.

**The OpenMP Common Core Memory Model**

A memory model defines the set of rules you must follow so you can safely overlap reads and writes to shared variables from multiple threads. There are three components to a memory model: (1) restricting the ability of a compiler to move instructions around reads/writes to shared variables, (2) ordering constraints on the instructions executing among two or more threads, and (3) ways to make a thread's temporary copies of shared variables (e.g., values in cache) consistent with the values in memory (i.e., flush).

In the OpenMP Common Core, we use synchronization constructs that include the required flush operations on memory. Those flushes handle compiler ordering constraints and force needed updates to memory. Therefore, we express the memory model in the OpenMP Common Core with three simple rules:

- When mixing reads and writes to shared variables from multiple threads, make sure there is a barrier (either an explicit barrier or an implicit barrier) between the writes and the subsequent reads.

- When updating shared variables inside a region where the control flow of the algorithm does not support a barrier, protect the update with a critical section.

- Never use the values of variables, even if they are updated in a race-free manner, to define an ordering constraint between threads. Ordering constraints are defined in the Common Core through barriers.

## 8.3 Working with Shared Memory

Consider the program shown in Figure 8.3. This program carries out an iterative computation. We break out of the loop once a convergence parameter is met or if the maximum number of iterations is exceeded. We do not show the code for the function `doit()`. For the sake of this example, assume it carries out an embarrassingly parallel operation over a distinct subset of elements from A where that subset is selected by the thread ID. We track the number of iterations by incrementing a shared variable and exit the while loop if the number of iterations exceeds some maximum value.

```
1
2   #include <omp.h>
3   #define TOL   0.0001
4   #define MAX   100000
5   #define NMAX 1000
6
7   //embarrassingly parallel computation, returns a convergence parameter
8   double doit(double *A, int N, int id);
9
10  int main()
11  {
12      int iter = 0;
13      int N = 1000;
14      double A[NMAX] = {0.0};
15      double conv=0.0;
16      #pragma omp parallel shared(A,N,iter) firstprivate(conv)
17      {
18          int id = omp_get_thread_num();
19          int nthrd = omp_get_num_threads();
20
21          while (iter < MAX) {
22              conv = doit(A, N, id);
23              if (conv < TOL) break;
24              if (id == 0) iter++;
25          }
26      } //end parallel region
27  }
```

Figure 8.3: **An erroneous program where updates may not be fully shared**
– This program carries out an iterative computation over the elements of an array **A**. Assume the function **doit()** carries out a computation that is embarrassingly parallel with a fixed subset of the array **A** selected by the thread ID. This program could fall into an infinite loop if the value of **conv** does not issue the break from the while loop and the shared variable **iter** is not propagated across all the threads allowing it to trigger the loop exit condition (**iter < MAX**).

The problem is that the updated value of **iter** may not be propagated to the other threads. A compiler does not usually take the execution of other threads into account when generating code. It is likely that a compiler would notice that **iter** is simply incremented during each pass of the loop and therefore there is no reason to go to a cache line to fetch a new value of **iter**. It could just keep its value of **iter** in a register. Without a flush, this program has no way to force all threads to see the updated values of **iter**. The fix to this problem is to make **iter** a **firstprivate** variable and have each thread update its private copy of the variable.

Another common problem occurs when using the nowait clause to turn off the barrier implied at the end of a worksharing-loop construct. Consider the code in Figure 8.4. In this program, a function is called inside the loop and the result is accumulated into a reduction variable. As we discussed earlier, a barrier is an expensive synchronization construct. If a subset of the threads takes much longer than the others to finish their assigned loop iterations, threads can wait at the end of the loop and greatly increase the parallel overhead. The solution is to add a nowait clause on the **parallel for** construct, but only if the program does not use the loop results later in the parallel region. This appears to be the case so the programmer places a nowait clause on the worksharing-loop construct in Figure 8.4. The problem is that the reduction operation to produce the final result in the variable **sum** is not guaranteed to be complete until the first barrier following the loop. Even if it was complete, without a flush there is no assurance that the value of **sum** is consistent among all threads. The solution is to remove the nowait or perhaps to put an explicit barrier before the function **another_job()** using **sum** is called.

```
1   #pragma omp parallel shared(A, B, sum)
2   {
3      int id = omp_get_thread_num();
4      int nthrds = omp_get_num_threads();
5      #pragma omp for reduction(+:sum) nowait
6         for (int i = 0; i < N; i++) {
7            sum += big_job(A,N);
8         }
9      bigger_job(B, id);      // a function that does not use A
10     another_job(sum, id);   // sum may not be available
11  }
```

Figure 8.4: **Reductions need a barrier** – This program carries out a computation inside a parallel loop and accumulates the result with a reduction. The function called after the loop uses the SPMD pattern and does not use any of the values computed in the loop, hence the programmer used a nowait clause. The last function uses the reduction variable which may not be available for all threads since the reduction is only guaranteed to complete at the next barrier following the loop. As a result, this is an incorrect program.

The memory model we have covered for the OpenMP Common Core is relatively simple. If you restrict yourself to the Common Core directives and are careful to avoid data races, this simplified memory model will let you safely write a wide range of shared memory algorithms. The Common Core, however, is quite limited and excludes a number of important patterns of synchronization. In particular, the contents of the OpenMP Common Core does not support synchronization between pairs of threads. The only synchronization constructs in the Common Core are collective synchronization constructs that work across the team of threads.

As an example of pairwise synchronization, consider the simple producer-consumer program in Figure 8.5. We first make sure that the runtime system gave us at least two threads. If not, then we exit the program. If we have two or more threads, we can continue into the program. One thread (the producer) will call a function to produce a result in the array A. The other thread (the consumer) will wait until the producer is finished at which point it will call a function to consume the result.

```
1   int flag = 0;  // a flag to communicate when the consumer can start
2
3   #pragma omp parallel shared(A, B, flag)
4   {
5       int id = omp_get_thread_num();
6       int nthrds = omp_get_num_threads();
7
8       // we need two or more threads for this program
9       if ((id == 0) && (nthrds < 2)) exit(-1);
10
11      if (id == 0) {
12          produce(A);
13          flag = 1;
14      }
15      if (id == 1) {
16          while (flag == 0) {
17              // spin through the loop waiting for flag to change
18          }
19      consume (A);
20      }
21  }
```

Figure 8.5: **Pairwise synchronization** – A producer-consumer pattern with one thread producing a result that another thread will consume. This program uses a spin-lock to make the consumer wait for the producer to finish. *Note: This program is not properly synchronized and as written will not work.*

This program provides a classic example of *pairwise synchronization*. Two specific threads need to enforce an ordering constraint between them, that is, one thread must finish a function before the next one can begin. A common way to do this is with a *spin-lock*. A shared variable is used as a flag to signal a condition between the two threads. It is set to an initial value (zero on this case) and the consumer waits until the variable changes value. It does this by spinning in a while loop until the variable changes.

The general concept is straightforward as seen in Figure 8.5. However, this program is incorrect and requires features from OpenMP that go well beyond the Common Core. You need a flush to assure that both threads see a consistent value for the flag variable. The flush we have discussed, however, is not a synchronization construct. It causes individual threads to make their view of memory consistent but it does not impose an order constraint between the threads. There are additional problems with the program in that the operation of storing the flag variable and later on reading it is not atomic. An atomic operation is uninterruptible and either completes or does not execute at all. On some computers, for variables that fit in a single machine word, basic loads and stores are atomic by default. This is the case, for example, with the x86 architecture. It is not the case in general, however, and therefore programmers should not depend on atomicity by default for loads and stores.

The need for pairwise synchronization is rare and most programmers go their entire career without ever needing to write such code. That is great since getting the right flushes and atomic constructs in place to make these types of program run correctly is difficult and seriously error-prone. For those curious about how these more advanced features work, we will revisit this problem in the next part of the book when we explore features of OpenMP that go beyond the Common Core.

## 8.4   Closing Comments

In this chapter we have introduced the issues raised by multiple threads running in a shared address space. Any time multiple threads are reading and writing the same locations in memory, it is possible to write programs which race to update variables and the resulting programs are undefined. These data races are very dangerous since the compiler in most cases cannot detect the error. The program will run and produce results that may look reasonable. It might even pass your suite of correctness tests. The data race, however, means the program is undefined and might sporadically produce answers that could be catastrophically wrong.

Therefore, programmers must be very careful to write programs that do not contain data races. This is much easier to do if you restrict yourself to the simplified memory model defined by the Common Core. While many important algorithms cannot be addressed by the Common Core subset of OpenMP, it turns out that for most application programmers working with parallel computers, this strategy is effective.

If you need pairwise synchronization or complex concurrent data structures where the synchronization protocols are integrated into the data structures themselves, you will need the full power of OpenMP. You will also need a great deal of expertise in the subject of concurrent algorithms. In all seriousness, most application programmers should probably never make such attempts. If at all possible, whenever you need to go beyond the memory model defined in the Common Core, find an expert to help you.

# 9 Common Core Recap

We have now covered the directives, clauses, and library routines that make up the OpenMP Common Core. Our journey progressed in five stages:

- **Threads and the OpenMP Programming Model**: We learned about threads and how to manage them. This included forking a team of threads, synchronization, and joining a team of threads. We also learned about the SPMD design pattern.

- **Parallel Loops**: We covered the worksharing-loop construct which is used to split-up work defined by a loop among a team of threads. A big part of working with loop-level parallelism is controlling how threads are scheduled for execution and the use of reductions.

- **OpenMP Data Environment**: The data environment is the set of variables visible within a region. We learned about the default rules for how variables move into and out of a data environment and a set of clauses we can use to modify those rules.

- **Tasks in OpenMP**: Tasks greatly expand the algorithms that can be addressed with OpenMP. We learned about tasks, the data environment for tasks, and the divide and conquer design pattern often used with tasks. We also addressed a second worksharing construct, the *single* construct.

- **OpenMP Memory Model**: What values can be legally returned when threads are reading and writing overlapping sets of variables? That's what a memory model is, and we learned about the simplified memory model we use with the OpenMP Common Core.

Five blocks of topics organized into five chapters that cover the Open Common Core. Our approach with this material was focused on pedagogy. We presented topics broken down into small chunks introduced in an order that supports effective learning. This is the right way to present material if the goal is to help people learn. It is a terrible way to present material if you want a reference guide: a place where you can go and quickly look up items from the Common Core.

In this chapter, the last chapter in our presentation of the OpenMP Common Core, we provide that reference guide. We assume you have read the previous five chapters and do not need us to describe any OpenMP concepts. You just need a

place to look up the detailed syntax and basic semantics of the OpenMP Common
Core to support your programming efforts.

## 9.1   Managing Threads

A thread encounters a `parallel` construct and creates a team of threads. Each
thread executes the code in the structured block that immediately follows the
parallel directive plus code in any functions called inside the structured block. The
full set of code that runs inside the parallel construct is called a *parallel region*.

The parallel construct is the only way to create threads in OpenMP. We summarize
the syntax of the construct and the available clauses in Table 9.1.

The size of the team of threads forked by a parallel construct is given by the
*nthreads-var* internal control variable[1]. The runtime system, however, may give you
fewer threads than requested. Once formed, the size of the team remains fixed until
the end of the parallel region. The thread that encountered the parallel construct is
part of the team. It is called the *master* thread with a thread ID equal to zero.

Table 9.1: **The parallel construct in C/C++ and Fortran** – This construct
forms a team of threads and starts parallel execution. Clauses included in the Common
Core are listed.

| |
|---|
| **#pragma omp parallel** *[clause[[,] clause]...]  new-line* <br>     structured block |
| **!$omp parallel** *[clause[[,] clause]...]* <br>     structured block <br> **!$omp end parallel** |

| |
|---|
| **shared** *(list)* <br> **private** *(list)* <br> **firstprivate** *(list)* <br> **default**(none) <br> **reduction** *(operator:list)* |

---

[1] An Internal Control Variable or ICV is an object maintained by the OpenMP runtime system.
It is "opaque" which means the actual object is hidden from the user. We use the concept of an
ICV to describe how features of the OpenMP system change as a program runs.

## 9.2   Worksharing Constructs

A worksharing construct is encountered by all the threads in a team. It divides the work in the structured block following the directive among the team of threads. A worksharing construct implies a barrier at the end of the construct. This barrier can be turned off with a `nowait` clause.

There are two worksharing constructs in the OpenMP Common Core:

- the *worksharing-loop* construct

- the *single* construct

The worksharing-loop construct is shown in Table 9.2. The worksharing-loop directive is followed by a loop. The iterations of that loop will be divided among the team of threads. The loop construct as used in the OpenMP Common Core uses the following canonical form for the `for` loop:

```
for (init-expr; test-expr; incr-expr)
    structured block
```

Basically, a compiler must be able to use the expressions in the loop control structure to build the logic for dividing blocks of loop iterations among threads. It is required that all threads in a team encounter the worksharing-loop construct and they must see the same values for any variables used in the loop control structures. Note that the loop control index for the loop parallelized by a worksharing-loop construct will be made into a private variable. There are two schedule kinds in the Common Core:

- **static**: Iterations are divided into chunks of size *chunk_size* and assigned to threads in the team in round-robin fashion in order of thread number.

- **dynamic**: Each thread executes a chunk of iterations then requests another chunk until none remain.

The other worksharing construct included in the Common Core is the *single* construct as shown in Table 9.3. All threads in the team must encounter the single construct. One thread executes the code in the region defined by the structured block associated with the construct. The other threads wait at the barrier implied by the end of the construct, unless this barrier has been disabled by the `nowait` clause.

Table 9.2: **The worksharing-loop construct in C/C++ and Fortran** – This construct specifies that the iterations of associated loops will be executed in parallel by threads in the team. The worksharing-loop construct is a worksharing construct. It has an implicit barrier unless turned off with **nowait**. Clauses included in the Common Core are listed.

| |
|---|
| **#pragma omp for** *[clause[[,] clause]. . . ]  new-line* <br>     for-loops |
| **!$omp do** *[clause[[,] clause]. . . ]* <br>     do-loops <br> **!$omp end do** *[nowait]* |

| |
|---|
| **private** *(list)* <br> **firstprivate** *(list)* <br> **nowait**                        (C/C++) <br> **reduction** *(operator:list)* <br> **schedule** *(kind[, chunk_size])* |

Table 9.3: **The single construct in C/C++ and Fortran** – This construct specifies that the associated structured block is executed by only one of the threads in the team. It has an implicit barrier unless turned off with **nowait**. Clauses included in the Common Core are listed.

| |
|---|
| **#pragma omp single** *[clause[[,] clause]. . . ]  new-line* <br>     structured block |
| **!$omp single** *[clause[[,] clause]. . . ]* <br>     structured block <br> **!$omp end single** *[nowait]* |

| |
|---|
| **private** *(list)* <br> **firstprivate** *(list)* <br> **nowait**                (C/C++) |

## 9.3   Parallel Worksharing-Loop Combined Construct

OpenMP defines a number of cases where two constructs can be combined into a single construct the semantics of which are identical to invoking the two constructs one after the other. In the OpenMP Common Core, we have one of these *combined constructs*: the parallel worksharing-loop construct described in Table 9.4.

Table 9.4: **The combined parallel worksharing-loop construct in C/C++ and Fortran** – This combined construct specifies a **parallel** construct containing one worksharing-loop construct with one or more associated loops. Clauses in Common Core are listed. It accepts any clauses that are accepted by the **parallel** or **for/do** directives, except the **nowait** clause.

| |
|---|
| **#pragma omp parallel for** *[clause[[,] clause]. . . ] new-line* <br>     for-loops |
| **!\$omp parallel do** *[clause[[,] clause]. . . ]* <br>     do-loops <br> **!\$omp end parallel do** |

| |
|---|
| **shared** *(list)* <br> **private** *(list)* <br> **firstprivate** *(list)* <br> **reduction** *(operator:list)* <br> **schedule** *(kind[, chunk_size])* |

## 9.4 OpenMP Tasks

Tasks provide a more flexible way to express parallelism in OpenMP. They support a wide range of irregular algorithms. The construct that creates an *explicit* task is defined in Table 9.5. A thread that encounters the task construct packages up the code for the task and the associated data environment into a task. The thread may opt to execute the task immediately, or defer execution. Essentially, if the execution is deferred, the task is placed in a work-queue. Available threads will work on the tasks in the queue until the queues are empty.

The common approach with tasks is to have one thread create tasks inside a single construct while the other threads, waiting at the barrier implied by the end of the single construct, work on the task work-queue. This pattern with a single construct, however, is not required and there are cases where all the threads in the team will generate explicit tasks in parallel. Tasks may be nested, that is, tasks may create tasks. This is actually quite common and is the basis of many recursive algorithms.

An important concept when working with tasks is to understand when tasks complete. OpenMP requires that threads will not exit a barrier until all tasks associated with that particular team of threads have completed. This includes all tasks created in the lexical scope of the construct that includes the barrier, and any tasks created by those tasks.

Table 9.5: **The task construct in C/C++ and Fortran** – This construct defines an explicit task. The data environment of the task is created according to data-sharing attribute clauses on **task** construct and any defaults that apply. Clauses included in the Common Core are listed.

| |
|---|
| **#pragma omp task** *[clause[[,] clause]... ] new-line* <br>     structured block |
| **!$omp task** *[clause[[,] clause]... ]* <br>     structured block <br> **!$omp end task** |

| |
|---|
| **shared** *(list)* <br> **private** *(list)* <br> **firstprivate** *(list)* <br> **default(none)** |

You can also wait for a subset of tasks to complete. In Table 9.6 we describe the `taskwait` directive. This directive causes the encountering thread to wait until its child tasks (i.e., the tasks created within the lexical scope of the `taskwait` directive) have completed[2]. This is a synchronization directive in that it defines an ordering relation between tasks.

Table 9.6: **The taskwait directive in C/C++ and Fortran** – This directive specifies a wait on the completion of the child tasks of the task that encounters the taskwait. The taskwait directive is a synchronization directive.

| |
|---|
| **#pragma omp taskwait** *new-line* |
| **!$omp taskwait** |

## 9.5   Synchronization and Memory Consistency Models

OpenMP is a multithreaded programming model. Threads execute in a shared address space. They are concurrent meaning they are not ordered with respect to each other. We exploit this concurrency to run the threads in parallel.

---

[2]The jargon can be complicated when mixing threads and tasks while explaining constructs in OpenMP. Technically, a thread runs a task and it is the "encountering task" that waits on its child tasks. We explain this task-oriented perspective for OpenMP in Chapter 13.

There are times, however, when we need to impose an order on operations by threads. These operations are called synchronization operations. We have already encountered one synchronization directive. This is the `taskwait` directive used to define ordering relations between sibling tasks (i.e. explicit tasks created in the lexical scope of the same parent task). We define two additional synchronization operations: `barrier` and `critical`. These support collective synchronization. Their function is defined in terms of behavior relative to the full team.

The `barrier` directive is encountered by all threads in the team. It defines a point in a region where all threads must wait until all other threads in the team arrive before any can continue. We describe this directive in Table 9.7.

Table 9.7: **The barrier directive in C/C++ and Fortran** – This directive specifies an explicit barrier at the point at which the directive appears. The barrier directive is a synchronization directive.

| |
|---|
| **#pragma omp barrier** *new-line* |
| **!$omp barrier** |

The final synchronization operation included in the OpenMP Common Core is the `critical` construct. This implements a *mutual exclusion* synchronization operation. Only one thread at a time can execute the structured block associated with the construct. If a thread encounters a critical construct and another thread is currently executing the "critical section", it will wait until the thread has completed executing the critical section. We describe this directive in Table 9.8.

Table 9.8: **The critical construct in C/C++ and Fortran** – This construct restricts execution of the associated structured block to a single thread at a time. The critical construct is a synchronization construct.

| |
|---|
| **#pragma omp critical** *new-line* |
| structured block |
| **!$omp critical** |
| structured block |
| **!$omp end critical** |

Synchronization operations are important in OpenMP. They prevent threads from conflicting when the logic of a parallel algorithm requires them to execute code in a particular order. They have an additional function. They support the memory consistency model of OpenMP.

Memory is a collection of variables where a variable is just a name we use to refer to an address in memory. If multiple threads mix loads and stores to an address, and if those loads and stores are not constrained to occur in a specified order, we have a data race. A program with a data race is ambiguous and therefore not legal.

A memory consistency model defines the rules needed to prevent data races. The restricted synchronization operations included with the OpenMP Common Core leads to a very simple memory consistency model. It is built around the concept of a *flush*. The operation of a flush is defined in terms of an individual thread. It causes that thread to make its own view of the shared address space consistent with memory. Write buffers and caches are flushed to memory. Cache lines being read are marked as "dirty" so the processor will need to refresh them from memory. The flush included in the Common Core is not a synchronization operation. It does not create an ordering relation among two or more threads. The flush is critical, however, to support synchronization. OpenMP implies a flush at certain points in OpenMP. These are summarized in Table 9.9.

Table 9.9: **Memory consistency rules** – Memory consistency rules define which values are allowed to be observed when a variable shared among two or more threads is read.

| **Memory Consistency Rules** |
| --- |
| A thread uses a flush to make its variables consistent with memory. A flush is implied at the following locations:<br>• Entry to and exit from the critical construct<br>• Exit from explicit and implied barriers |

## 9.6   Data Environment Clauses

A data environment is the set of variables visible within a region. These variables have a data sharing attribute: *shared* if they are in memory visible to the team of threads or *private* if they are only visible to a single thread.

When a thread (or a task) encounters a construct that creates a region, we need to understand how variables visible to the encountering thread (or task) move into the created region. Variables in the data environment of the encountering thread (or task) are called the *original variable*. The data environment clauses in Table 9.10 modify the default rules for how original variables interact with the newly created region.

Table 9.10: **The data sharing clauses** – The data sharing attribute clauses apply only to variables whose names are visible in the construct on which the clause appears. **shared**: the items in the list are shared among threads or explicit tasks executing the construct. **private**: creates a new variable for each item in the list that is private to each thread or explicit task. The private variable is not given an initial value. **firstprivate**: declares list items to be private to each thread or explicit task and assigns them the value the original variable has at the time the construct is encountered.

| |
|---|
| **shared** *(list)* |
| **private** *(list)* |
| **firstprivate** *(list)* |

## 9.7   The Reduction Clause

In the OpenMP Common Core, we support the operation of reduction into a scalar. The reduction clause is defined in Table 9.11. For each variable in the list of a reduction clause, the following occurs:

- Create a private variable with the same name as the variable from the list in the clause.

- Initialize that private variable to the identity of the operator (also known as the *reduction-identifier*).

- The region executes normally with the private variable created by the reduction.

- At the end of the region, combine the private variables from the list according to the operator from the reduction clause.

- Combine the reduced value computed at the end of the region with the original variable using the operator from the reduction clause.

Table 9.11: **The reduction clause** – The reduction clause specifies an operator and one or more list items.

| reduction(operator:list) | |
|---|---|

| Operator | Initialization Value |
|:---:|:---|
| + | 0 |
| * | 1 |
| - | 0 |
| min | Largest representable number in **reduction** list item type |
| max | Smallest representable number in **reduction** list item type |

## 9.8    Environment Variables and Runtime Library Routines

Most of the action with OpenMP happens when a program is compiled. The directives provide explicit commands to the compiler which creates multithreaded code in response to them. There are actions in OpenMP, however, that do not happen at compile time. They can only happen as a program runs. These actions occur through runtime library routines or through environment variables.

We will start with library routines and environment variables that interact with the Internal Control Variable for the default number of threads (*nthreads-var*) to be forked by a `parallel` construct. This variable has an implementation dependent default value. It can be set when a program is launched through the value of an environment variable `OMP_NUM_THREADS` (see Table 9.12). A function from the OpenMP runtime library routine can *set* the variable during the program's execution (`omp_set_num_threads()` shown in Table 9.13).

Table 9.12: **The runtime environment variable OMP_NUM_THREADS** – This runtime environment variable sets default number of threads to request for parallel regions.

| **OMP_NUM_THREADS** list |
|---|

The next pair of functions help us understand the threads inside a parallel region. For many algorithms, you need to know the number of threads in a team and the thread rank for each thread in the team where the *rank* ranges from 0 (for the master thread of the team) up to the number of threads minus 1. To find the number of threads in a team, call the `omp_get_num_threads()` function described in Table 9.14.

Table 9.13: **The omp_set_num_threads runtime library routine in C/C++ and Fortran** – This runtime function sets the default number of threads to request for subsequent parallel regions. It is illegal to call this function within the dynamic extent of a parallel region.

| |
|---|
| void **omp_set_num_threads**(int *num_threads*); |
| **subroutine omp_set_num_threads**(*num_threads*) <br> integer *num_threads* |

To return the rank for each thread in the team, call the omp_get_thread_num() function described in Table 9.15.

Table 9.14: **The omp_get_num_threads runtime library routine in C/C++ and Fortran** – This runtime function returns the number of threads in the current team of the innermost enclosing parallel region.

| |
|---|
| int **omp_get_num_threads**(void); |
| integer **function omp_get_num_threads**() |

Table 9.15: **The omp_get_thread_num runtime library routine in C/C++ and Fortran** – This runtime function returns the thread number of the calling thread within the current team of the innermost enclosing parallel region.

| |
|---|
| int **omp_get_thread_num**(void); |
| integer **function omp_get_thread_num**() |

The final runtime library routine included in the OpenMP Common Core is omp_get_wtime() shown in Table 9.16. This function returns a `double` value which represents the time in seconds from some fixed point of time in the past. This "time" is specifically *wallclock time* meaning it is the time you would see on a standalone clock independent of the computer (e.g., such as a clock on your wall). If you call this function twice, before the start of a block of code being timed and after that block of code has completed, the difference between the two values will be the elapsed time for executing that block of code.

Table 9.16: **The omp_get_wtime runtime library routine in C/C++ and Fortran** – This runtime function returns elapsed wall clock time in seconds. It is not guaranteed to be globally consistent across all the threads.

| |
|---|
| double **omp_get_wtime**(void); |
| double precision **function omp_get_wtime**() |

# III  BEYOND THE COMMON CORE

The OpenMP Common Core is a great place to start. Many OpenMP programmers remain within the OpenMP Common Core their entire careers. A well-rounded OpenMP programmer, however, needs to know how to go beyond the Common Core when situations demand it. This part of the book will help you take those first steps.

We start with Chapter 10 which remains grounded in multithreading for hardware you can still approximate as an SMP. We discuss clauses missing from the Common Core used with the `parallel`, `worksharing-loop` and `task` constructs. We then look at functionality we skipped altogether when we defined the Common Core.

Chapter 11 brings us back to the memory consistency model for OpenMP. Quite bluntly, this is the most difficult topic in all of shared memory programming. Programmers should avoid writing software that depends on the more subtle aspects of memory models. There are times, however, when you have no choice. A good example is synchronization between pairs of threads; the so-called *pairwise synchronization* problem. We explore this and other advanced topics in synchronization. More importantly, however, we lay the foundation for understanding the latest memory model developments as expressed in the OpenMP 5.0 standard.

In Chapter 12, we survey a range of topics as we look at hardware beyond the SMP systems addressed by the Common Core. We start with a detailed study of Non-Uniform-Memory-Access (NUMA) architectures. Given the fact that most SMP systems are actually NUMA systems (an idea we explore at length) the techniques we address should be used much more often than they are. We discuss the vector units in CPUs and how to program them with OpenMP using the various SIMD directives. The approach we take is perhaps a bit unusual. Rather than walk through the long list of OpenMP SIMD directives, we instead focus on a step by step transformation of a program into highly vectorized code. Our goal is to give you a deep understanding for how vectorization works. Finally in this complex chapter on hardware, we take a look at the GPU and how to program it from OpenMP.

We close the book with an important chapter on what you can do to continue your education in OpenMP. We build the case that books cannot keep up with the pace of OpenMP evolution. That means eventually you will need to read the OpenMP specifications. Given that eventuality, we use our time in Chapter 13 to go back over the fundamental structure of OpenMP; but this time we use (and explain) the formalized jargon found in the specification. The goal is that you can go from our final chapter directly into the OpenMP specification to explore the full breadth of the language.

# 10 Multithreading beyond the Common Core

Deciding what to include in the OpenMP Common Core was in part motivated by pedagogy. In making our decisions, we consistently favored those constructs needed to learn the core ideas behind OpenMP and understand the most common ways it is used. Three core constructs lay the foundation for the Common Core: `parallel`, the *worksharing-loop*, and explicit tasks. The rest of the Common Core supports these constructs by controlling the data environment, managing the threads in a team, and defining constraints on the order of operations (for example, to prevent data races).

In this chapter, we survey the elements of OpenMP that were *not* included in the OpenMP Common Core but are still frequently used. We break the chapter down into two parts. We start with the constructs in the common core (`parallel`, `task`, and the *worksharing-loop*) and the more important clauses used with those constructs. Then we describe directives that expand on the multithreading functionality provided by the OpenMP Common Core.

Throughout the chapter, we continue with our focus on pedagogy. We cover the most commonly used elements of the directives and the concepts needed to understand them deeply. As we have stated repeatedly, this book is not a reference guide. Therefore, for any given directive, clause, or construct, there may still be details we do not cover. In each case, a more complete discussion can be found in the book *Using OpenMP – The Next Step* [13].

## 10.1   Additional Clauses for OpenMP Common Core Constructs

Multithreading is where OpenMP began. While we have expanded well beyond basic multithreading over the last decade, it still remains the "heart and soul" of OpenMP. In the Common Core we covered the foundational constructs of multithreading; the `parallel` construct to create threads, the *worksharing-loop* construct to split loop iterations among threads, and the `task` construct to create explicit tasks. These constructs and their data environments can be modified using a number of clauses. We covered the most commonly used clauses as part of the Common Core. In this section, we present other clauses that are often used though not quite so often as the Common Core clauses.

### 10.1.1  The Parallel Construct

The parallel construct is used to create a new team of threads. We have used it extensively in this book. In Table 10.1 we show the parallel construct with the most frequently used clauses. We have covered many of these clauses already. Focusing on the clauses introduced here, we start with `num_threads`. This clause takes an integer expression to set the number of threads to request for the parallel region. For example, if we have some maximum number of threads to work with (`MaxThreads`) and we want to use a quarter of them in one parallel construct, we would use the directive:

```
#pragma omp parallel num_threads(MaxThreads/4)
```

The number of threads requested by the `num_threads` clause supersedes the default number of threads (based on the *nthreads-var* ICV[1]) but only for the parallel region associated with the clause. Successive parallel regions will by default request a number of threads equal to the *nthreads-var* ICV.

Managing threads for a parallel region can add considerable overhead to a program. Forking threads, either by creating them anew or pulling them from a thread-pool, can consume thousands of CPU cycles. The join at the end of the parallel region can consume an additional batch of thousands of cycles. If the logic in a program suggests that the work inside a parallel region is too small to justify incurring the thread management overheads, you may want to skip creation of the team and instead execute the region on a single thread. This is done with the `if`(*scalar-expression*) clause. If the scalar expression evaluates to true (in C, any value other than 0 is true), the team of threads is created. If false, the team is not created and the parallel region executes with a single thread. For example, when nesting parallel regions inside other parallel regions, you must assure that the total number of threads used across all parallel regions does not grow so large that the Operating System wastes inordinate amounts of time managing too many threads relative to the number of available cores (the so-called *oversubscription problem*). The following code might be used to mitigate this situation.

```
int nthreads = omp_get_num_threads();
#pragma omp parallel if(nthreads < MaxThreads/4) num_threads(4)
```

---

[1] The initialism ICV expands to "Internal Control Variable". An ICV is used by the OpenMP runtime to manage default values or to control features of the system as a program runs.

Table 10.1: **A parallel construct in C/C++ and Fortran** – This construct is used to create a team of threads. The most commonly used clauses are listed.

| |
|---|
| **#pragma omp parallel** *[clause[[,] clause]...] new-line*<br>    structured block |
| **!$omp parallel** *[clause[[,] clause]...]*<br>    structured block<br>**!$omp end parallel** |

| | |
|---|---|
| **if** *(scalar-expression)* | (C/C++) |
| **if** *(scalar-logical-expression)* | (Fortran) |
| **shared** *(list)* | |
| **private** *(list)* | |
| **default(shared \| none)** | (C/C++) |
| **default(shared \| firstprivate \| private \| none)** | (Fortran) |
| **firstprivate** *(list)* | |
| **reduction** *(operator:list)* | |
| **num_threads** *(integer-expression)* | (C/C++) |
| **num_threads** *(scalar-integer-expression)* | (Fortran) |
| **copyin** *(list)* | |
| **proc_bind (master \| close \| spread)** | |

The `default` clause was introduced when we discussed the OpenMP data environment. Use of `default(none)` tells the compiler that the storage attribute of every variable declared prior to the parallel construct and used inside the parallel region must be explicitly defined. You can also declare that variables declared prior to the parallel region are `shared`, which is the default behavior. Fortran defines two other default cases: `private` and `firstprivate`. These are not available in C/C++ since it is possible to present a named item to the compiler that appears to be a variable but resolves at runtime to a constant. Hence, it is not possible for a compiler to create a private copy of such an item for each thread at compile time.

There are two remaining clauses in Table 10.1: the `copyin` and `proc_bind` clauses. We will discuss `copyin` later in this chapter when we discuss the `threadprivate` directive and discuss `proc_bind` in Chapter 12 when we discuss NUMA systems.

As an example of how these newly introduced clauses are used, consider the program in Figure 10.1. A unitary (trace preserving) transform is applied to a matrix. The details of the matrices or the transformation process itself is not relevant for this example so the functions `initMats()` and `transform()` are not shown.

Within the parallel region, `transform()` is set up for use of the SPMD pattern with the `id` and `Nthrds` passed in as arguments. This is followed by the trace computation (the sum of the diagonal elements) with a worksharing-loop construct.

We use `default(none)` to force the compiler to flag any cases where a variable used inside the parallel region but declared prior to the `parallel` directive is not defined in a data environment clause. We request that 4 threads be used for this parallel region with the `num_threads(4)` clause.

Finally, notice the `if(N>100)` clause. This clause indicates that if the matrix order, N, is greater than `100`, the parallel region should be executed with multiple threads. If this condition is not met, the parallel region will be executed with a single thread. This is an important capability in OpenMP and should be used to make sure thread management overhead is not incurred for problems too small to benefit from multiple threads.

### 10.1.2   The Worksharing-Loop Construct

To many programmers, OpenMP is a technology for creating threads and mapping the iterations of loops among those threads. This was definitely the mindset when we created OpenMP 1.0, and while the language has grown considerably, sharing the work of a loop among a team of threads remains one of the most important capabilities in OpenMP.

The OpenMP worksharing-loop construct and its most commonly used clauses are described in Table 10.2. Most of these clauses are already familiar since they are included in the Common Core. We added one additional data environment clause, `lastprivate`. As with any of the data environment clauses, `lastprivate` takes as an argument a comma-separated list of variables. These variables are shared among the team of threads that encounter the worksharing-loop construct. As with `private` and `firstprivate`, the `lastprivate` clause directs the compiler to create a private copy of each variable in the list. At the end of the worksharing region, the original variable for each of the variables in the lastprivate list will be assigned the value from the last iteration of the loop where by "last" we mean the last iteration as defined by a sequential execution of the loop. For example, in the following simple loop:

```
1   #include <stdio.h>
2   #include <stdlib.h>
3   #include <omp.h>
4
5   // initialization and transform functions
6   // (we will not show the function bodies)
7   extern void initMats(int N, float *A, float *T);
8   extern void transform(int N, int id, int Nthrds, float *A, float *T);
9
10  int main(int argc, char**argv)
11  {
12      float trace=0;
13      int i, id, N, Nthrds;
14      float *A, *T;
15
16      // set matrix order N
17      if (argc == 2)
18          N = atoi(argv[1]);
19      else
20          N = 10;
21
22      // allocate space for three N x N matrices and initialize them
23      T = (float *) malloc(N*N*sizeof(float));
24      A = (float *) malloc(N*N*sizeof(float));
25      initMats(N, A, T);
26
27      #pragma omp parallel if(N>100) num_threads(4) default(none) \
28                  shared(A,T,N) private (i,id,Nthrds) reduction(+:trace)
29      {
30          id = omp_get_thread_num();
31          Nthrds = omp_get_num_threads();
32          transform(N, id, Nthrds, T, A);
33
34          // compute trace of A matrix
35          // i.e., the sum of diagonal elements
36          #pragma omp for
37          for (i = 0; i < N; i++)
38              trace += *(A+i*N+i);
39      }
40      printf(" transform complete with trace = \%f\n",trace);
41  }
```

Figure 10.1: **Examples of clauses on the Parallel construct** – The matrix A is transformed by a transformation which is assumed to be a unitary transform (i.e., a trace preserving transform). Notice how continuation of a pragma onto an additional line is indicated by a backslash. We do not show code for initMats() and transform() as their function bodies are not relevant for this example.

```
#pragma omp for lastprivate(ierr)
    for (int i = 0; i < N; i++)
        ierr = work(i);
```

the value of `ierr` assigned to the original variable at the end of the region will be the value from whichever thread executed the last iteration of the loop (iteration `i=N-1`). It is illegal for a single variable to appear in more than one data environment clause, with one exception: OpenMP allows you to put the same variables in the `firstprivate` and `lastprivate` clauses. In other words, the OpenMP specification understands that regardless of whether or not your algorithm needs the value of a variable from the last iteration of a loop, you still may need to provide an initial value for a private variable.

Table 10.2: **A loop worksharing construct in C/C++ and Fortran** – The construct used to divide the work defined by a loop among a team of threads including the most commonly used clauses. The omitted clauses address SIMD execution, schedule modifiers, and the ordered clause (used for synchronization and doacross loops).

| **#pragma omp for** *[clause[[,] clause]...]* |
|---|
| for-loops |
| **!$omp do** *[clause[[,] clause]...]* |
| do-loops |
| **!$omp end do**[nowait] |

| |
|---|
| **private** *(list)* |
| **firstprivate** *(list)* |
| **lastprivate** *(list)* |
| **reduction** *(operator:list)* |
| **schedule** *(kind [, chunk_size])* |
| **collapse** *(n)* |
| **nowait**                                    C/C++ |

The `schedule` clause takes an argument indicating the kind of schedule followed by an optional parameter for the size of the block of iterations used as a unit of scheduling (the so-called `chunk_size`). The following schedule *kinds* are defined.

- **static**: The loop iterations are divided into chunks of size *chunk_size* and are assigned to the threads in the team in a round-robin fashion (i.e., as if dealing out a deck of cards). If a *chunk_size* is not indicated, the implementation will provide one chunk of iterations per thread but the number of iterations per thread is not specified.

- **dynamic**: The loop iterations are divided into chunks of size *chunk_size*. Each thread executes a chunk and then requests an additional chunk. If a *chunk_size* is not provided, the default *chunk_size* is one.

- **guided**: An alternative form of the dynamic schedule where the *chunk_size* starts as some large value and is reduced each time a new chunk of iterations executes until the minimum size of *chunk_size* is reached. This is done to reduce runtime overhead of managing the scheduling of chunks.

- **auto**: The compiler and the runtime can schedule the iterations of the loop as they choose. It does not have to be any of the other defined *kinds*.

- **runtime**: The schedule and potentially the *chunk_size* are taken from the internal control variable *run-sched-var*.

The `schedule(runtime)` clause is useful for changing the schedule on a loop without recompiling the program. The internal control variable, *run-sched-var*, is set to the value of the string defined by the environment variable, `OMP_SCHEDULE`, when the OpenMP program begins execution. For example, with the Bash shell command line interpreter and prior to running the program you could type:

```
export OMP_SCHEDULE="dynamic,7"
```

This causes any worksharing-loops in the program that have the `schedule(runtime)` clause to use a dynamic schedule with *chunksize=7*. Different schedules could be explored by just changing the value of the `OMP_SCHEDULE` variable.

As with most internal-control variables, the schedule and chunk size can also be manipulated by runtime library routines. These functions are described in Figure 10.2. The functions are paired, "`omp_set`" and "`omp_get`", to set or query the value of the internal control variable. The schedule *kind* is defined by a C enum type as shown in Figure 10.2 so you can set the *omp-schedule* corresponding to the case above with the statement

```
omp_set_schedule(omp_sched_dynamic,7);
```

An example of the runtime schedule clause and its supporting functions is shown in Figure 10.3. This program is taken from a simple molecular dynamics program that simulates the melting of a periodic crystal of argon. By periodic we mean

```
// runtime schedule function in C/C++
void omp_set_schedule (omp_sched_t kind, int chunk_size);
void omp_get_schedule (omp_sched_t* kind, int* chunk_size);

typedef enum omp_sched_t {
  omp_sched_static = 1,
  omp_sched_dynamic = 2,
  omp_sched_guided = 3,
  omp_sched_auto = 4
} omp_sched_t;

!$ Runtime schedule functions in Fortran

subroutine omp_set_schedule(kind, chunk_size)
integer (kind=omp_sched_kind) kind
integer chunk_size

subroutine omp_get_schedule(kind, chunk_size)
integer (kind=omp_sched_kind) kind
integer chunk_size

integer(kind=omp_sched_kind), parameter::omp_sched_static=1
integer(kind=omp_sched_kind), parameter::omp_sched_dynamic=2
integer(kind=omp_sched_kind), parameter::omp_sched_guided=3
integer(kind=omp_sched_kind), parameter::omp_sched_auto=4
```

Figure 10.2: **Support for manipulating the schedule at runtime** –Library routines and associated data structures to set and query the runtime schedule Internal Control Variable in C/C++ and Fortran. While OpenMP defines the values within the enumerated type for the schedules, it is good programming practice to use the names, not the values.

that if the atomic coordinates fall outside a boundary, we wrap it around to the opposite boundary. This is an instance of an N-body problem with each atom of argon dependent on the position of every other atom. The calculation scales as the number of atoms squared and quickly becomes untenable for even modest sized crystals.

Various means are used in N-body problems to circumvent this N-squared problem. In this program, we show the "cutoff" method. The potential that determines the force on each atom depends on the separation distance between each pair of atoms. Electrostatic forces, of course, drop off rapidly with distance. Hence, we can pick a distance (the cutoff distance) and assume any atoms further away than the cutoff distance do not significantly contribute to the force.

We have a number of points we wish to make with this example. First, notice the loop structure on lines 15 and 21. The outer loop runs over each atom in the crystal, but the inner loop runs over atoms with labels greater than the outer loop control index. This works since the force between atoms i and j is equal and opposite to the force of atom j and atom i, so we accumulate the force on atom j inside the innermost loop and then set the opposite of that force on atom j. This loop structure, combined with the physical dependence on the distance of pairs of atoms with respect to the cutoff, means that the work per atom on the outer loop varies and is unpredictable. This is a classic instance of where the *dynamic* or even the *guided* loop schedule is useful. Experimenting with a range of values is straight forward and avoids recompiling for each case by using the `schedule(runtime)` clause and the `OMP_SCHEDULE` environment variable to explore the different options.

Near the end of the code in Figure 10.3, we show an example of how to query and then print schedule information. This is inside an `#ifdef` block since in a production run with the program, printing runtime schedule information for each step through the molecular dynamics code would be overwhelming. On line 44 we define a variable of type `omp_sched_t` to hold the value of the enum type for the schedule kind. We fetch the schedule kind and the chunk size with the call to `omp_get_schedule` on line 45. On line 46, we print the kind and chunk size. Notice the trick we use to print a string corresponding to the enum type for kind. The OpenMP specification defines the actual values of the kind enums so we can build an array of strings on line 7 and use the enum values to reference the right string. Since OpenMP does not define an enum value of zero, the string for that value indicates an error.

Regardless of the schedule, a program should not depend on any particular mapping of loop iterations onto specific threads. By design, implementations are given considerable freedom in deciding how best to carry out that mapping. There is one exception to this rule. If the following conditions are met, you can depend on

```
 1  #include <omp.h>
 2  #include <stdio.h>
 3
 4  #define DEBUG 1
 5
 6  // map schedule kind enum values to strings for printing
 7  static char* schdKind[] = { "ERR"," static "," dynamic "," guided "," auto"};
 8
 9  // external function for potential energy term
10  extern double pot(double dist);
11
12  void forces(int npart,double x[],double f[],double side,double rcoff)
13  {
14     #pragma omp parallel for schedule(runtime)
15        for (int i = 0; i < npart*3; i += 3) {
16
17        // zero force components on particle i
18        double fxi = 0.0; double fyi = 0.0; double fzi = 0.0;
19
20        // loop over all particles with index > i
21           for (int j = i + 3; j < npart * 3; j += 3) {
22
23              // compute distance between i and j with wraparound
24              double xx = x[i] - x[j];
25              double yy = x[i+1] - x[j+1];
26              double zz = x[i+2] - x[j+2];
27
28              if(xx<(-0.5*side)) xx+=side; if(xx>(0.5*side)) xx-=side;
29              if(yy<(-0.5*side)) yy+=side; if(yy>(0.5*side)) yy-=side;
30              if(zz<(-0.5*side)) zz+=side; if(zz>(0.5*side)) zz-=side;
31              double rd = xx * xx + yy * yy + zz * zz;
32
33              // if distance is inside cutoff radius, compute forces
34              if (rd <= rcoff*rcoff) {
35              double fcomp = pot(rd);
36              fxi += xx*fcomp;    fyi += yy*fcomp;    fzi += zz*fcomp;
37              f[j] -= xx*fcomp;  f[j+1] -= yy*fcomp; f[j+2] -= zz*fcomp;
38              }
39           }
40        // update forces on particle i
41           f[i] += fxi;   f[i+1] += fyi;   f[i+2] += fzi;
42        }
43     #ifdef DEBUG
44        omp_sched_t kind;
45        int chunk_size;
46        omp_get_schedule(&kind, &chunk_size);
47        printf("schedule(%s,%d)\n",schdKind[kind],chunk_size);
48     #endif
49  }
```

Figure 10.3: **Use of runtime schedules** – Function computes forces in a simple molecular dynamics program. Prints information about the runtime schedule when enabled by the DEBUG macro.

the mapping between loop iterations and threads to be consistent between different loops:

- The `static` schedule is explicitly used and the chunk size is the same between loops.

- The loops in question have the same number of loop iterations.

- The loops bind to the same parallel region (i.e., they are executed by the same team of threads).

Depending on the same mapping between loop iterations and a team of threads can be useful if data within the caches associated with different threads can be utilized between the loops. It can also be used when there are fixed patterns of dependencies inside the two loops so correct results would be produced when a `nowait` clause is used on the earlier loop.

The clause `collapse(n)`, where `n` is a positive integer expression, defines the number of loops to associate with the worksharing-loop construct. It stipulates that the `n` loops immediately following the worksharing-loop construct are to be combined into a single implied "super-loop". This implied loop has a much larger iteration space and runs through the loop iterations in the order that would result from a serial execution of the program. Any additional clauses, such as data environment clauses or reductions, are applied to the larger implied loop.

The loops that are combined each must follow the standard rules for a legal worksharing-loop. The loops are usually "perfectly nested" by which we mean there is no intervening code between the loops (only with OpenMP 5.0 are non-perfectly-nested loops allowed and the rules for supporting them are complicated). The `collapse` construct is used to increase the size of a parallel loop to create enough work to effectively balance the load across the team of threads.

We show an example of the `collapse` clause in Figure 10.4. The `Apply()` function applies an input function (`MFUNC`) to each element of a two-dimensional array. We want this function to be able to efficiently handle any size of the array dimensions, `N` and `M`. We use an `if` clause on the `parallel for` construct to force the system to use only one thread for small values of `N` and `M` (i.e., when their product is 100 or less). For large values of `N`, parallelizing the single outer loop would probably be effective, but for smaller values, there may not be enough loop iterations to maintain a well-balanced load among the threads or to overcome the overheads of managing the parallel loop. Hence, we use the `collapse(2)` clause to tell the compiler to

```
1   #include <omp.h>
2
3   // apply a function (*MFUNC) to each element of an N by M array
4
5   void Apply(int N, int M, float* A, void(*MFUNC)(int, int, float*))
6   {
7      #pragma omp parallel for num_threads(4) collapse(2) if(N*M>100)
8         for (int i = 0; i < N; i++)
9            for (int j = 0; j < M; j++)
10              MFUNC(i, j, (A+i*M+j));
11  }
```

Figure 10.4: **The collapse clause on a worksharing-loop construct** – The `Apply` function applies a function input as a function pointer to each element of an `N` by `M` array, `A`. Note that the pointer expression (`A+i*M+j`) points to the (`i,j`) element of the array `A`.

create an implicit loop that run over `N*M` loop iterations. This larger loop should hopefully have enough loop iterations to support efficient execution of this code.

There are a number of features of the worksharing-loop construct that we have not discussed. The `simd` clause will be discussed later when we discuss mapping loop iterations onto the vector units of a processor. Others (`ordered` and the schedule modifiers) are relatively new and not widely used. To learn more about these clauses, consult the book *Using OpenMP – the Next Step* [13].

### 10.1.3   The Task Construct

Tasks greatly expand the range of algorithms that can be supported by OpenMP. The task construct and the clauses commonly used with `task` are shown in Table 10.3. The data environment clauses (`private`, `firstprivate` and `shared`) are part of the Common Core. The `default` clause on a task construct works the same as it does on the `parallel` construct. For the `if` clause, if the integer expression (or logical-expression in the case of Fortran) evaluates to false, then the task is immediately executed by the thread that encountered the task construct (i.e., task execution is not deferred).

The `untied` clause is a bit more complicated to describe. Consider the execution of an OpenMP program based on explicit tasks. One or more tasks create tasks and fill a task queue with deferred tasks waiting to execute. Some threads in the

Table 10.3: **A task construct in C/C++ and Fortran** – The construct used to create an explicit task and a subset of the most commonly used clauses.

| |
|---|
| **#pragma omp task** *[clause[[,] clause]...]* <br>     structured-block |
| **!$omp task** *[clause[[,] clause]...]* <br>     structured-block <br> **!$omp end task** |

| | |
|---|---|
| **if** *(scalar-expression)* | (C/C++) |
| **if** *(scalar-logical-expression)* | (Fortran) |
| **shared** *(list)* | |
| **private** *(list)* | |
| **firstprivate** *(list)* | |
| **default(shared \| none)** | (C/C++) |
| **default(shared \| firstprivate \| private \| none)** | (Fortran) |
| **untied** | |
| **priority** *(priority-value)* | |
| **depend** *(dependence-type : list)* | |

team execute tasks that create new tasks and others execute tasks waiting in the task queue. If the task queue grows too large, the OpenMP runtime system might pause tasks that are generating new tasks to free up threads to work on tasks in the queue. At the same time, some tasks may be blocked waiting at a `taskwait` synchronization point. A runtime might pause those waiting tasks so the blocked thread can work on tasks waiting in the queue. In these cases, the runtime system switches a task from active execution into a suspended state.

The OpenMP standard defines when task switching is allowed to occur. This can occur at *task scheduling points*. These are points in the execution of a task when it checks in with the runtime system to see if task switching might be indicated. The task scheduling points are task creation, task completion, task wait, and barriers. OpenMP supports an additional, explicit task scheduling point called `taskyield`. Programmers can place a `#pragma omp taskyield` in their code at points where they suspect task switching can help keep threads busy doing productive work, such as when a task has a significant chance of causing a thread to block while waiting for some resource.

When a task switches from a suspended state to an active state, a thread picks up the work in the thread at the statement right after the task scheduling point. A task is said to be *tied* if the system guarantees that when the work resumes with that task, the same thread that worked on the task before suspension would work with the task *after* suspension. It is important for a task to be tied to a thread if there are resources local to that thread a task might depend on (such as `threadprivate` data discussed later in this chapter). A task is *untied* if the thread that executes the task after it has suspended at a task scheduling point can be different than the thread that executed the task prior to suspension.

Defining a task as untied can have large performance benefits when a program is frequently switching tasks. The safest state for a task, however, is to be *tied* since a programmer may not be aware of the ways a task is implicitly depending on resources local to a thread. Therefore, the default state for a task is *tied*. A programmer must explicitly mark a task as untied with the `untied` clause when the task is created.

Up to this point, we have discussed two ways to control the order tasks execute: the `taskwait` synchronization directive (pause while waiting from tasks in the lexical extent of the taskwait to complete) and the barrier (pause until all deferred tasks have completed). Algorithms often need more fine-grained control over the order tasks execute. For example, in some cases there are clear priorities between tasks. The semantics of the algorithm may not require a particular ordering of tasks, but for performance reasons you want some tasks to run before others. You can indicate this preference with the `priority` clause. This clause is a hint to the OpenMP runtime system about the priority of execution for the task generated by a task construct. The priority-value is a non-negative integer expression. For the set of deferred tasks waiting to execute, the tasks with a greater priority-value are *recommended* to execute before those with a lower priority-value.

The priority-value ranges from 0 to a maximum value indicated by the *max-task-priority-var* Internal control variable (ICV). By default, the max-task-priority-var ICV is zero. It can be set by through the environment variable `OMP_MAX_TASK_PRIORITY`. For example, to set the maximum priority to 50 under the Bash shell, you would use the following commands:

```
export OMP_MAX_TASK_PRIORITY=50
```

To query the value of the `OMP_MAX_TASK_PRIORITY` ICV, you can use the function:

```
int omp_get_max_task_priority(void);
```

Unlike the other ICVs, there is no runtime library routine to set *max-task-priority-var*. Given the dynamic nature of tasks and the task queues that manage them, it is not clear how to safely allow the maximum priority value to vary dynamically, so it is set once when the program starts through the `OMP_MAX_TASK_PRIORITY` environment variable.

Priorities are hints. They suggest an order for tasks, but do not force the system to follow that order. When explicit orders between tasks are demanded by the algorithm, we need a mechanism to define execution orders between tasks. We do this through the *depend* clause. Before we explain the semantics of the depend clause, let's consider when dependencies might be needed in an OpenMP algorithm. A frequent way to think about algorithms is in terms of a Directed Acyclic Graph (DAG). We show a simple DAG in Figure 10.5. The elliptical elements labeled with letters indicate computations and the directed arcs between the nodes are dependencies between the nodes. DAGs are heavily used in the machine learning community and many shared memory dense linear algebra libraries are built around a DAG where the nodes are calls to BLAS (Basic Linear Algebra Subprograms).

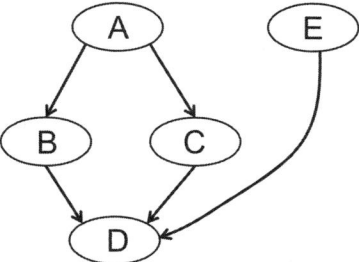

Figure 10.5: **A simple Directed Acyclic Graph where the dependencies between nodes are indicated by directed arcs** – The elliptical elements labeled with letters represent nodes with computations.

We build a DAG in an OpenMP program using tasks with the **depend** clause. The depend clause specifies a *dependence-type* and a comma separated list of variables which may be scalar variables or array sections. These variables must be in scope (i.e., visible) to the sibling tasks involved in the DAG. The allowed dependence types are *in*, *out*, or *inout*. A variable with the *in* dependence type causes the task to wait for a task to complete that has the same variable in a **depend** clause with the *out* dependence type. In other words, tasks with paired depend clauses over the same variables causes the task with the *in* depend clause to wait until the task with the corresponding *out* depend clause to complete. Recall from Chapter 8 that task creation and task completion imply a flush; hence thread visible variables in a task with the *depend(out)* clause are flushed to memory before the task with the *depend(in)* clause begins to execute, which of course will execute a flush to memory before it begins execution. Therefore, memory consistency between tasks connected by a dependency arc is managed "automatically" on behalf of the programmer. The last remaining dependence-type is *inout*. The *inout* dependence-type indicates variables used both as an *in* dependence to control when a task begins and as an *out* dependence to control when subsequent sibling tasks execute.

We provide in Figure 10.6 an example of a program implementing the DAG from Figure 10.5. One thread executes the code in the single region to create the deferred tasks, one for each node in the DAG. The other threads in the team wait at the barrier implied at the end of the single construct and execute the tasks within the task queue. The arcs in the DAG are easy to read from the connections between *depend(out)* and *depend(in)* clauses.

The task constructs are an important part of OpenMP. While we have covered the most commonly used elements of OpenMP for tasks, there is much more. OpenMP has added ways to map loop iterations onto tasks (the *taskloop* construct). It has also added a new task synchronization construct to cover child and sibling tasks (the *taskgroup* construct). To improve the efficiency of task programs, additional clauses control when to directly execute a task (*final*), and when to assume the data environment of a task can be shared with a child task (*mergeable*). To learn more about tasks, consult the book *Using OpenMP – The Next Step* [13] or the resources described in Chapter 13.

## 10.2   Multithreading Functionality Missing from the Common Core

The directives and clauses included in the Common Core capture the most commonly used elements of OpenMP. They were also selected with an eye to effective pedagogy; that is they are the right place to start when learning about multithreaded programing. There are a few elements of OpenMP that are extremely important for a well-rounded programmer working with threads to know. They were left out of the Common Core since they complicate learning the foundational skills of programming

```
1   #include <omp.h>
2
3   //functions Awork through Ework not shown
4
5   int main()
6   {
7       float A, B, C, D, E;
8       #pragma omp parallel shared(A, B, C, D, E)
9       {
10          #pragma omp single
11          {
12              #pragma omp task depend(out:A)
13                  Awork(&A);
14              #pragma omp task depend(out:E)
15                  Ework(&E);
16              #pragma omp task depend(in:A) depend(out:B)
17                  Bwork(&B);
18              #pragma omp task depend(in:A) depend(out:C)
19                  Cwork(&C);
20              #pragma omp task depend(in:B,C,E)
21                  Dwork(&E);
22          }
23      }
24  }
```

Figure 10.6: **Task Dependencies** – This program implements the DAG shown in Figure 10.5. The functions represent the nodes and the edges of the DAG are captured by the patterns of **depend** clauses.

with threads, but they still need to be understood before your basic education in multithreaded programming with OpenMP is complete.

We cover these vital topics in this section. This includes data that is private to a thread but shared inside that thread (**threadprivate**), isolating work to the master of a team (**master**), operations that read and write memory without interruptions (**atomic**), and controlling how much memory is set aside for each thread (**OMP_STACKSIZE**).

### 10.2.1   Threadprivate

The basic memory model of OpenMP views memory as a set of variables that give names to addresses in memory. We have worked with two types of variables: **shared** and **private**. OpenMP defines a third type of memory: **threadprivate**.

Threadprivate memory is private to a thread. It cannot be accessed by other threads. The variables in threadprivate memory, however, are scoped by the host language to have visibility across routines. Informally, you can think of threadprivate memory as private to a thread but global inside the thread.

We define the threadprivate directive in Table 10.4. It is a declarative directive which means it appears in a program where variables are declared and impacts the semantics of their declaration. The global scope variable (or named common block in Fortran) is declared in the host language and specified with the `threadprivate` directive to place the listed variables (or common blocks) in threadprivate memory.

Table 10.4: **Threadprivate directive in C/C++ and Fortran** – For C/C++, *list* is a comma separated list of file-scope, namespace-scope, or static block-scope variables that do not have incomplete types. For Fortran, *list* is a comma-separated list of named variables and named common blocks where the common block names appear between slashes.

| |
|---|
| **#pragma omp threadprivate***(list)* |
| **!$omp threadprivate***(list)* |

Threadprivate variables are initialized according to the rules of the host language. In many cases, the initialized value is specified when the variable that will become threadprivate is first declared in the host language. Initialization happens once and occurs at some unspecified point in the program execution before the variable is first used. We provide a program that uses `threadprivate` in Figure 10.7. The program is adapted from the linked list program in Figure 7.9 from Chapter 7. The list is traversed in the `while` loop (lines 23 to 30) and some work is carried out for each node in the list (`processwork(p)` on line 27). A single thread traverses the list and creates a task (line 24) for each node in the list.

We define a counter on line 5 and initialize it to zero. This counter is made `threadprivate` on line 6. A function, `inc_count()`, increments the counter each time it is called. Since a copy of `counter` has been replicated for each thread, as each thread calls `inc_count()` on line 26, we keep track of how many tasks each thread handled. The full set of tasks complete at the barrier at the end of the single construct (line 31) after which we print the thread ID and the value of `counter`. This is a common use of threadprivate variables in OpenMP. Note that this technique for counting tasks assigned to each thread would not work if the tasks were untied.

```
1    #include <stdio.h>
2    #include <sys/time.h>
3    #include <omp.h>
4
5    int counter = 0;
6    #pragma omp threadprivate(counter)
7
8    void inc_count()
9    {
10       counter++;
11   }
12
13   int main()
14   {
15       p = init_list(p);
16       head = p;
17
18       #pragma omp parallel
19       {
20           #pragma omp single
21           {
22               p = head;
23               while (p) {
24                   #pragma omp task firstprivate(p)
25                   {
26                       inc_count();
27                       processwork(p);
28                   }
29                   p = p->next;
30               }
31           }
32           printf("thread \%d ran \%d tasks\n",omp_get_thread_num(),counter);
33       }
34       freeList(p);
35
36       return 0;
37   }
```

Figure 10.7: **Counting task executions with a threadprivate counter** – This program traverses a linked list in parallel with tasks doing a random amount of work for each node in the list. A threadprivate variable is used to keep track of how many tasks were executed by each thread. Note: we do not provide the functions used for the list nor the list processing.

This approach to counting tasks assigned to threads translates directly into Fortran. The use of the threadprivate construct in Fortran is sufficiently different from that

in C/C++ that we provide an example of a counter in Fortran in Figure 10.8 (taken
from page 211 of the OpenMP 4.5 Examples document). The threadprivate variable
itself (`COUNTER`) is placed in a named common block (`/INC_COMMON/`). The common
block name placed between slashes appears in the threadprivate directive.

```
1          INTEGER FUNCTION INCREMENT_COUNTER()
2          COMMON/INC_COMMON/COUNTER
3  !$OMP  THREADPRIVATE(/INC_COMMON/)
4          COUNTER = COUNTER +1
5          INCREMENT_COUNTER = COUNTER
6          RETURN
7          END FUNCTION INCREMENT_COUNTER
```

Figure 10.8: **A counter with threadprivate in Fortran** – This code come from
the OpenMP 4.5 Examples document (threadprivate.1.f). This Fortran function uses the
same logic as our previous example with C. You create a global scope variable in Fortran
through common blocks. Hence, the counter is placed in a named common block and that
block is made threadprivate.

Initialization is typically carried out by static declarations defined inside the
program source code. For more dynamic cases where you want program logic
to determine the values to initialize threadprivate data at runtime, you can use
the `copyin(list)` clause on a `parallel` construct. The values from the original
variables (i.e., the variables with the same name prior to `parallel` construct) are
copied into the threadprivate variables once at some point before they are first
used inside the parallel region. The `copyin` clause is one of the few cases where a
threadprivate variable is allowed to appear in a clause on an OpenMP construct.

Threadprivate data exposes details of the threads used in an OpenMP program.
This is a subtle but important point. Typically in an OpenMP program, you do not
care which specific thread runs which loop iteration or handles any particular task.
Threadprivate data, however, is tied to specific threads and hence introduces sources
of error in a program. Any program where a thread accesses the threadprivate
data of another thread is *nonconforming*; i.e., it is erroneous but the OpenMP
specification cannot define it as an *error* since there is no way to guarantee that a
compiler or runtime can detect this situation. Another source of error is assumptions
about threadprivate data persisting between parallel regions. This works but only
in cases where: (1) the number of threads does not change between parallel regions

(a concept called *static mode* that we will discuss later), and (2) the parallel regions are not nested inside other parallel regions.

### 10.2.2  Master

The `master` construct defines a block of work that is carried out by the master thread of the team. The syntax of the `master` construct is shown in Table 10.5. Unlike the `single` construct, it does not imply a barrier at the end of the construct. The master thread does the work in the structured block and the other threads continue to execute the statements following the `master` construct.

Table 10.5: **The Master construct in C/C++ and Fortran** – The structured block associated with this construct will be executed by the master thread of the team.

| |
|---|
| **#pragma omp master** |
| structured block |
| **!$omp master** |
| structured block |
| **!$omp end master** |

The `master` construct is logically equivalent to calling a runtime library routine to get the thread number followed by an if statement to single out the thread associated with `id == 0`:

```
int id = omp_get_thread_num();
if (id == 0) {
    structured-block
}
```

It may seem to be strange to include this construct given how easy it is to represent the functionality of `master` with an existing library routine. We included `master` in OpenMP, however, since it expressed through a pragma what would otherwise require multiple lines of executable code. Recall that a pragma is ignored if the compiler does not recognize the pragma. Therefore, by expressing this functionality through a pragma, the code will work consistently with compilers that support OpenMP and those that do not.

### 10.2.3   Atomic

The `atomic` construct ensures that a variable (i.e., a specific storage location in memory) is read, written or updated as a distinct, uninterrupted action. The `atomic` construct protects a variable from the possibility of multiple, simultaneous updates to a storage location by concurrent threads, which would result in a data race. The `atomic` construct shares a great deal in common with a critical section in that the atomic operation occurs with mutual exclusion. If multiple threads try to execute an `atomic` construct at the same time, the "first thread" will carry out the atomic operation while the other threads will wait their turn. The `atomic` construct, however, is less general than a `critical` construct. It was designed around the atomic operations included in the instruction sets of modern processors and is potentially more efficient than a critical construct.

The `atomic` construct is defined in Table 10.6. A clause defines the type of atomic operation of which we describe only the three most common cases: *read*, *write*, and *update* (we do not include *capture*). The default case (i.e., the case when no clause is included) is *update*. Each type of atomic construct applies to a different form of expression-statement that follows the atomic directive.

Table 10.6: **The atomic construct in C/C++ and Fortran** – This construct is used to support uninterruptible memory operations. We include the most commonly used clauses with the construct. The variables `x` and `v` are scalar variables that you can assign to (i.e., l-values) and `binop` is one of `+`, `*`, `-` plus the bitwise and shift operators. `expr` is an expression that does not include the variable `x`.

| **#pragma omp atomic** [*clause*] *new-line* |
|---|
| *expression-stmt* |
| **!$omp atomic** [*clause*] |
| *expression-stmt* |
| **!$omp end atomic** |

| clause | expression-stmt |
|---|---|
| **read** | `v = x;` |
| **write** | `x = expr;` |
| **update** | `x++; x--; ++x; --x;` |
| *the default case* | `x binop= expr; x = x binop expr;` |
|  | `x = expr binop x;` |

As a simple example of the atomic construct, consider the numerical integration program from Chapter 4. In the version of the program to introduce synchronization constructs in OpenMP (Figure 4.11), we used a `critical` section to safely update the sum with the partial sum from each thread:

```
#pragma omp critical
    full_sum += partial_sum;
```

We could do this same operation with an `atomic` construct, possibly reducing overhead since the atomic construct is designed to map onto a single hardware instruction. Looking at Table 10.6, we see that the type of atomic construct corresponding to the statement in the critical region is the *update*. This is the default case for atomic so we can replace the `critical` in Figure 4.11 with:

```
#pragma omp atomic
    full_sum += partial_sum;
```

While similar to a `critical` section, it is important to appreciate that the `atomic` construct only applies to the operation that directly involves the storage location in memory. For example, consider the following `atomic` construct:

```
#pragma omp atomic
    full_sum += foo();
```

The execution of the function `foo()` is not protected by the `atomic` construct. It is executed by the system "as if" by the following statements:

```
tmp = foo();
#pragma omp atomic
    full_sum += tmp;
```

Any side-effects inside `foo()` would be exposed to potential data races; unlike the analogous case with a critical construct.

We have barely scratched the surface of the full capabilities of atomics in OpenMP and the many variations defined in the specification. To go deeper into atomic, we need to understand it in the context of the detailed memory consistency model of OpenMP. Hence, we defer this discussion until Chapter 11.

### 10.2.4   OMP_STACKSIZE

OpenMP was designed to support a wide variety of systems. Hence, the specification avoids details that pertain to the operating system and the threads that support

OpenMP. There is one detail of these threads, however, that we can not avoid: the stack associated with each thread.

A process is managed by an operating system on behalf of an executing program. The process forks threads which remain associated with the process. When the operating system creates the threads, it sets aside some local memory for each thread. This memory is managed as a stack. While global memory and memory shared among the threads reside in the heap associated with the process, private variables reside in a thread's stack.

The stack has a finite size and if code running inside a thread creates large objects in memory (such as arrays), the thread's stack memory can overflow leading to potentially catastrophic failure. To deal with this problem, OpenMP defined an internal control variable called *stacksize-var*. It controls the size of the memory stack associated with each thread in the team. This variable can only be set once when a program begins execution. It is set through an environment variable called OMP_STACKSIZE. For the Bash shell, the command to set the *stacksize-var* ICV is as follows:

```
export OMP_STACKSIZE=20000
```

The value of the stacksize-var takes a positive integer and a letter suffix to indicate the units for the stack size. The units defined by OpenMP includes:

- *size* sets the size in units of KiloBytes

- *size*B sets the size in units of Bytes

- *size*K sets the size in units of 1024 Bytes

- *size*M sets the size in units of 1024 KiloBytes

- *size*G sets the size in units of 1024 MegaBytes

The following set of examples (taken directly from the OpenMP specification) show how to work with the different units (assuming use of the bash shell):

```
export OMP_STACKSIZE="3000K"
export OMP_STACKSIZE="10M"
export OMP_STACKSIZE="10M"
export OMP_STACKSIZE="1G"
export OMP_STACKSIZE=20000      # this is in KiloBytes
```

How a system responds when there are problems with the stack size is difficult to define in a way that can be supported across a wide range of systems. Therefore, how a system responds when asked for more memory than can be provided or when the stack size is too small and a thread overflows its stack is undefined, but it tends to be rather nasty.

### 10.2.5  Runtime Library Routines

#### 10.2.5.1  omp_get_max_threads

You can ask the OpenMP runtime how many threads are in a team by using `omp_get_num_threads()` but only from inside a parallel region. There are times when you need a function you can call from outside a parallel region to find the maximum number of threads you might get in a team created by a subsequent `parallel` construct. For example, you might need to allocate memory for an array providing a buffer for each thread in a team. You would use the runtime library function from Table 10.7 to find this number.

Table 10.7: **Library routine to return the maximum number of threads in C/C++ and Fortran** – Unless overridden by a `num_threads` clause, the team created by a parallel region will not exceed this number.

| |
|---|
| int **omp_get_max_threads**(void) |
| integer **function omp_get_max_threads**() |

You can override the default number of threads to use in parallel region by the use of a `num_threads` clause or by a call to `omp_set_num_threads()`. In both cases, the maximum number of threads could change.

#### 10.2.5.2  omp_set_dynamic

An OpenMP program typically consists of multiple sequential parts separated by parallel regions. The OpenMP runtime can try to optimize the size of the team of threads from one parallel region to the next. This is called *dynamic mode*. Dynamic adjustment of the number of threads in a team can be important when the load on the system is highly variable; so over the course of the execution of a program, differing numbers of threads would allow more effective utilization of system resources.

Variation in the number of threads from one parallel region to the next, however, means the OpenMP runtime must assume resources associated with the threads may change between parallel regions. If you wish to reuse thread resources (such as threadprivate memory or loop schedules) between parallel regions, you need to tell the runtime system to turn off the capability of dynamic thread adjustment. Once dynamic execution is disabled, the runtime system is said to be in *static mode*.

Dynamic mode is enabled or disabled with a call to the function in Table 10.8.

Table 10.8: **Library routine to set the mode for a program in C/C++ and Fortran** – Allow team sizes to vary between parallel regions when the integer value `dyn_threads` is `true`. In C, any nonzero, integer value is true. In Fortran, pass the logical variable with value `.TRUE.` to the subroutine.

| |
|---|
| void **omp_set_dynamic**(int dyn_threads) |
| **subroutine omp_set_dynamic**(dyn_threads) <br> logical dyn_threads |

### 10.2.5.3   omp_in_parallel

Letting the number of active threads exceed the number of physical cores can compromise performance as the operating system consumes resources with excessive thread swapping. This is called *oversubscription*. Hence, there are times when you want to know if you are inside an active parallel region so you can adjust the number of threads created in subsequent parallel regions or skip creating new threads altogether. This is particularly useful when developing library routines, which typically means the context within which the routines will be called is unknown. With the `omp_in_parallel()` function, logic in your code can be used to decide to either fork additional parallel regions or not based on whether it was or was not called within a parallel region.

The function in Table 10.9 returns true (i.e., nonzero in C) if it is called inside an active parallel region.

Table 10.9: **Library routine to query if code is inside a parallel region** – The function returns *true* if it is called inside an active parallel region.

| |
|---|
| void **omp_in_parallel**(void) |
| logical **function omp_in_parallel**() |

## 10.3    Closing Comments

With this chapter, we began our exploration of OpenMP beyond the Common Core. We started with additional clauses used with the three Common Core constructs: parallel, worksharing-loop, and task. These clauses give us more control over the execution of these constructs and expand the range of algorithms we can address. For example, we added dependencies to the task construct which let us express Directed Acyclic Graphs with OpenMP.

We also considered a number of features in OpenMP not addressed by the Common Core constructs. We discussed the `master` construct which lets you conveniently define code that will be run by the master thread of a team. We covered atomics in OpenMP: a topic that will be extremely important when we explore enhancements to the OpenMP memory model added in later versions of OpenMP. We also described an additional storage class with the `threadprivate` directive which lets you define data that is private to a thread but global across functions run by a thread.

Finally, we considered some features of OpenMP that impact how the overall program executes. This includes static vs. dynamic modes which controls whether an implementation of OpenMP can vary the number of threads from one parallel region to the next. We also discussed ways to explicitly set the size of a thread's stack.

# 11 Synchronization and the OpenMP Memory Model

In Chapter 8 we described the memory model used in the OpenMP Common Core. This simplified model was used in the original versions of OpenMP. It is defined in terms of two basic operations: a `flush` and a `barrier`. The `flush` operation makes the temporary values of variables (e.g., values held in cache or a register file) consistent with those in memory (i.e., in RAM). The `barrier` operation is the fundamental synchronization operation in the Common Core defining a fixed point around which memory operations can be organized. It is a *collective synchronization* operation meaning that it applies an ordering constraint to all the threads in a team. This model works and is easy to use. In most cases, the flushes are implied and expressed through mutual exclusion operations (*critical constructs*), taskwait synchronization, and the barriers at the end of parallel regions and worksharing-loop constructs (unless turned off with a `nowait` clause).

This Common Core memory model is straightforward for programmers to use, but it is quite limited. There are cases where the flush/barrier based model adds too much overhead. For example, why define a synchronization point that orders memory operations for all threads in a team when all the algorithm might require is for a pair of threads to interact: the so-called *pairwise synchronization* problem. We alluded to this problem in Chapter 8 but we did not solve it. In this chapter, we will solve this problem. We will do so by introducing a more complex memory model defined in later versions of OpenMP. This model is aligned with modern programing language design principles and is based on *flush* operations and *atomic* operations.

Another problem with collective synchronization defined across the full team of threads is that it can only be expressed as directives inside an OpenMP region. There are cases, where you need a synchronization protocol to be incorporated into the definition of complex data structures. We address this case with an explicit, mutual exclusion operation called a *lock*.

Finally, C++ added threads to the core language in C++11. This required the addition of a formal memory model to C++. Other languages, such as C and OpenCL, have based their own memory models on the C++ memory model. With OpenMP 5.0, we have joined this trend by rewriting the OpenMP memory model to align it with the C++ programming language. Hence, a well-rounded parallel programmer needs to be aware of the memory model in modern C++ (i.e., C++11 and beyond). We will briefly touch on this model and how it maps onto OpenMP.

## 11.1   Memory Consistency Models

OpenMP threads execute within a shared memory: an address space accessible by all the threads in a team where variables are stored. The only way to make a shared memory system efficient is if the threads are allowed to maintain a temporary view of memory that resides in memory structures between the processor and the RAM. The details of these structures are not defined in the OpenMP specification, but they usually include register files, caches, and local write buffers.

In practical terms, a memory consistency model (or just "memory model" for short) defines the value that can be returned by an operation that loads a value from memory. In this case, the word "memory" refers to the shared memory accessible by all the threads. Hence, the crux of the matter comes down to how we manage the temporary views of memory for each thread with respect to the order of load/store operations by each of the threads.

When threads interact through variables in shared memory, they must make their temporary view of memory consistent with shared memory. They do this through a `flush` construct. In the OpenMP Common Core, we described flush as an operation that causes values in registers or those written into the cache hierarchy to be written to memory, cache lines to be marked invalid so the next time they are accessed they will load from memory, and whatever else is needed on a particular system to make the thread's view of memory consistent with that in shared memory. In the Common Core, we did not explain how to explicitly invoke a flush. We instead focused on constructs that implied a flush (such as task constructs and barriers).

To explore the full complexity of the OpenMP memory model, however, we need to discuss explicit flushes. The `flush` directive is defined in Table 11.1. The *flush-set* is the set of shared variables to which the flush applies. By default, the flush set is all the shared variables available to a thread and accessible at the time the thread encounters the `flush` directive. An optional clause to flush provides a comma-separated list of shared variables. These define a reduced flush set; letting a programmer define a subset of variables to flush.

In addition to addressing memory consistency, a flush also interacts with the rules governing when a compiler can reorder instructions. A compiler is not allowed to reorder instructions around a flush if the instructions use a variable from the flush set. This means that a programmer can analyze a flush and how it relates to instructions in program order to reason about the sharing of values in the flush set with other threads.

| **#pragma omp flush** *[(flush-set)]* *new-line* |
| **!$omp flush** *[(flush-set)]* |

Table 11.1: **A flush in C/C++ and Fortran** – An executable directive that causes the thread's view of shared memory to be consistent with shared memory. The optional *flush-set* is a comma-separated list of shared variables to which the flush is applied.

At this point, it would be useful to provide an example of how to use a flush. We cannot do this, however, since a flush used in isolation is a dangerous directive. This is part of the reason why in the OpenMP Common Core we did not support an explicit flush. We will defer examples of how to use a flush to the next section when we discuss the pairwise synchronization problem.

Returning to our central problem, namely to unambiguously define which values can be legally returned when a variable is loaded from memory, we need to discuss how a flush (or any other operation for that matter) is ordered in time with other operations executed by a program. For this discussion, we will adopt the traditional language used in memory model literature and talk about an instruction or when necessary the operations invoked by an instruction as an *event*. The foundational work on understanding the order of execution of events from concurrent threads comes from Leslie Lamport [6]. His formalization of how concurrent threads interact, established over 40 years ago, is used to this day. His approach did this without assuming an absolute, precise time-reference available to all threads. This work was inspired by the way time appears in the theory of special relativity where you can only be assured of the sequence of events between specific inertial reference frames, but you cannot assume an absolute ordering of events with respect to an external, absolute clock.

The crux of Lamport's work is the concept of concurrency organized around *happens-before relations*. The instructions from concurrent threads are unordered with respect to each other outside of specific synchronization events. There is no way in a scalable system to assign a time stamp to each instruction and put them in a fixed order in time. All you can do is establish a "happens-before" relationship; i.e., we can say that for a properly synchronized program, certain events happen before others.

What exactly does it mean to have a *happens-before* relation between events? We start by considering happens-before relations running on a single thread. To do this we need to introduce a new term: a sequenced-before relation.

- Sequenced-before: A partial order between events executed by a single thread. Given two events, A and B, we say that A is sequenced before B if evaluation of A (including any side-effects implied by A) completes before the execution of B.

The concept of *sequenced-before* is tightly coupled to a concept familiar to the language design community of a *sequence point*. Consider the statements that make up a program. These include declarations, arithmetic operations, function calls, memory allocations; basically, anything we use to build a program. Some of these statements, such as a simple assignment of a scalar, map directly onto the low level instructions of the computer. You execute a store operation and it completes without launching other operations. There are no side effects. Other statements, such as declaring and initializing an array, involve multiple low level operations, i.e., allocate memory, store a value, increment a pointer to the next location in memory, store another value, and so forth until the initialization is done. That single declaration and initialization statement launched a number of operations on the computer creating a range of possible side effects. In both the simple and complex cases, we describe the idea of a *sequence point* as the point in the program's execution when the statement in question is done and any side effects associated with the statement are complete.

An exhaustive list of the rules defining sequence points is not worthwhile for our purposes. We will just consider the most important rules and a few examples. In the C programming language, the most common sequence points include the following:

- The end of a full expression. This includes initializations, expressions within control flow statements such as if, while, and for, and expressions on return statements.

- At the logical operators && and ||, the conditional ?, and the comma operator.

- At a function call, in particular just after the evaluation of all arguments but just before the call.

- Immediately before a function returns.

There are other sequence points in C, but those listed above cover the most common cases and are enough for us to understand how they relate to memory consistency.

We provide some typical examples of sequence points in Figure 11.1. Each of the declarations on lines 2 to 4 are sequence points. The declaration on line 11 is a sequence point as well, but so are the subexpressions a=1, b=2 and c=0. Since the comma operator is a sequence point, these three subexpressions occur with a well-defined order: a=1 is sequenced-before b=2 which is sequenced-before c=0. The full statement on line 17 defines a sequence point. This statement contains two calls to functions. Since the + operator is not a sequence point, the function calls inside the expression on line 17 are not ordered. Hence we say that the sequence points for the function calls are *indeterminately sequenced*, i.e., they do not define a sequenced-before relation.

On lines 22 to 27, we have a for loop. The loop expression on line 22 is a sequence point that contains three sequence points, one for each subexpression. These are ordered expressions and hence define a sequenced-before relation between them. Inside the body of the for loop on line 26, there is an expression where a function is called with arguments that are themselves function calls. Each function call is a sequence point, however, in the C programming language, the order that arguments to a function are evaluated is not defined. Hence, this is once again a case of sequence points that are *indeterminately sequenced*.

As our final example from Figure 11.1, consider the expression on line 33. This is a subtle point in the C programming language. The full expression is a sequence point but it contains two operations: an assignment to the variable a and an increment of the variable a. This is a case where assignment to a can happen before the increment has completed. It is essentially a race condition with one thread. This case is not allowed and the subexpressions are *unsequenced*.

In summary, a sequence point is a point in the execution of a program where an event is complete as are any side effects associated with that event. We have three cases for the ordering of the sequence points in a program:

- Sequenced-before: A relationship between sequence points when one sequence point proceeds another.

- Indeterminately sequenced: A relationship between sequence points when they execute in some order but that order is not defined.

- Unsequenced: A condition that holds when sequence points conflict and lead to undefined results.

The happens-before relation in a single thread follows directly from the concept of sequenced-before relations. If the event A is sequenced-before the event B, then A *happens-before* B. If the event A is indeterminately sequenced with event B, they

```
1    // each declaration is a sequence point
2    extern int func1(int, int);
3    extern int func2(int);
4    extern int func3(int);
5
6    int main()
7    {
8    // The two comma operators plus the full expression define sequence
9    // points ... all ordered by sequenced−before relations.
10
11       int a = 1, b = 2, c = 0;
12
13   // 3 sequence points: the full statement plus the 2 function calls.
14   // The + operator is not a sequence point so the function calls are
15   // unordered and therefore, indeterminately sequenced.
16
17       d = func2(a) + func3(b);
18
19   // each expression in the for statement is a sequence point.
20   // they occur in a sequenced−before relation.
21
22       for (int i = 0; i < N; i++) {
23           // function invocations are each a sequence point. Argument
24           // evaluations are unordered or indeterminately sequenced.
25
26           func1(func2(a), func3(b));
27       }
28
29   // Mixing of a store and an increment on the same variable in the
30   // same statement. They are unordered and define a race
31   // condition. The increment and store are unsequenced.
32
33       a = a++;
34   }
```

Figure 11.1: **Examples of sequence points** – This program fragment demonstrates the most commonly encountered examples of sequence points and the relations sequenced-before, indeterminately sequenced, and unsequenced.

do not have a *happens-before* relationship. Defining a happens-before relation inside a single thread is trivial once you understand the sequence points defined by a programming language. Happens-before relationships become much more interesting when we consider statements in two or more concurrent threads.

To define a happens-before relation between threads, we must synchronize the threads. In the OpenMP Common Core, we defined synchronization in terms of collective synchronization operations that apply across the team of threads. We now want to use a more fine-grained synchronization concept, where we consider synchronization as a specific event among two or more threads. We do this by defining a *synchronized-with* relationship.

A synchronize-with relationship holds between threads when they coordinate execution around an event in order to define an ordering constraint to their execution. The concept is best understood through an example. Consider the execution of two threads shown in Table 11.2. We can view the execution of each thread as a sequence of events. Think of each event as a sequence point and further that each thread's events are ordered by sequenced-before relations. A special event (which will be described at length when we discuss the pairwise synchronization problem) is used to define a synchronized-with relation. In other words, the two threads coordinate their execution and order their sequence of events around those involved in the synchronized-with relation. This lets us define the two happens-before relations between threads:

- $A_1$ happens-before $B_1$ happens-before $S_1 \longleftrightarrow S_2$ happens-before $C_2$

- $A_2$ happens-before $B_2$ happens-before $S_2 \longleftrightarrow S_1$ happens-before $C_1$

Table 11.2: **Synchronizes-with and happens-before** – Two threads execute a sequence of events. On each of the threads the events $A_i$, $B_i$, $S_i$, and $C_i$ are ordered through sequenced-before relations. The events $S_1$ and $S_2$ are special events that define a synchronized-with relation between the two threads. From these, we can define happens-before relations between events on different threads.

| Thread 1 | Thread 2 |
|---|---|
| Event $A_1$ | Event $A_2$ |
| Event $B_1$ | |
| | Event $B_2$ |
| Event $S_1 \longleftrightarrow$ | Event $S_2$ |
| Event $C_1$ | Event $C_2$ |

We now have the tools we need to understand the order of events among multiple threads. We associate the general term "event" with sequence points. We understand how to order sequence points through sequenced-before relationships. To order events between threads we use synchronize-with relations. Putting the two concepts together, we can understand the happens-before relations among threads. By tracking the flushes, both implied flushes and explicit flushes, we can define values that can be read from memory at different points in a program (which is exactly what a memory model is designed to do).

We have covered a lot of ground without much in the way of example code. This was necessary since we really need sequence points, happens-before, flushes, and synchronized-with relations all together before we can move to code. We will do this in the next section where we discuss the pairwise synchronization problem.

## 11.2    Pairwise Synchronization

In Chapter 8 we observed that pairwise, or *point-to-point*, synchronization is not supported in the OpenMP Common Core. Now that we are working beyond the Common Core, we will revisit this problem and show how to solve it. We will drive this discussion by considering the producer-consumer pattern. This is a common pattern used in parallel computing. One thread, the producer, carries out some work to produce a result. The other thread, the consumer, waits until the producer is finished and then consumes the result. This serializes the processing between the two steps and hardly seems worthy of discussion in an exploration of parallel computing. This pattern, however, is very common inside what is commonly called *pipeline parallelism*.

We show a schematic representation of pipeline parallelism in Figure 11.2. In the top half of the figure we show producer and consumer steps as a linear sequence to be executed by a single thread. We create a processing pipeline by creating two threads: one to run the producer tasks and the other to run the consumer tasks. Once the pipeline is full, in this case, after the first producer task is done, the two threads can proceed in parallel. As long as the number of pipeline stages is large enough so the serial work at the beginning (filling the first pipeline stage) is not significant, the performance of these pipeline parallelism programs can be quite good.

The crux of pipeline parallelism is how a single producer-consumer pair of concurrent threads interacts. We show a program that implements a producer-consumer pair in Figure 11.3. It is important to stress that this program is incorrect. Over the course of this section, we will show how to make it correct.

This program uses a spin-lock to implement pairwise synchronization. A spin lock uses a simple variable as a flag to signal between threads. This must be visible

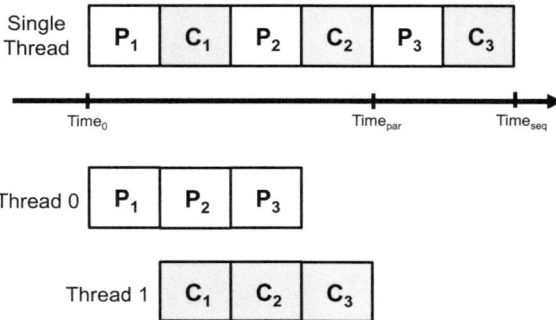

Figure 11.2: **Pipeline parallelism** –A process consisting of two stages: a producer stage ($P_i$) and a consumer stage ($C_i$). Executed sequentially (the upper part of the figure), the process takes $Time_{seq}$. By running on two threads with pipeline parallelism (the lower part of the figure), the two stages overlap once the pipeline has been filled dropping the run time to $Time_{par}$.

to both threads so we declare `flag` prior to creating the threads. We create the threads with a parallel construct on line 4 using a `shared(A, flag)` clause just to emphasize that these variables are shared among threads. After creating the team of threads, we check to make sure we actually got the two threads we asked for on line 10. Then we choose one thread (thread 0) to be the producer and the other thread (thread 1) to be the consumer.

The `flag` variable is initialized to zero when it was allocated. The producer does its work to produce the result, the variable `A`, and then sets `flag` to establish a synchronized-with relation with thread 1. Meanwhile on thread 1, the producer thread, executes the essence of the spin-lock on lines 17 to 19. This is a `while` loop that spins until the value of the variable `flag` becomes non-zero. At that point, it exits the loop and the event that completes its part of the synchronized-with relation. Thread 1, knowing that its synchronized-with event is sequenced-before the `consume(A)` in line 20, and that the `produce(A)` on thread 0 was sequenced-before the synchronized-with event on thread 0, can safely assume that it will load a consistent value of `A` to use in the `consume(A)` function.

The program in Figure 11.3 does not work. It fails for two reasons, and in explaining each of these reasons, we will cover the key issues you must understand when working with memory models and synchronization. There are two aspects

```
1   int flag = 0;   // a flag to communicate when the consumer can start
2   omp_set_num_threads(2);
3
4   #pragma omp parallel shared(A, flag)
5   {
6       int id = omp_get_thread_num();
7       int nthrds = omp_get_num_threads();
8
9       // we need two or more threads for this program
10      if ((id == 0) && (nthrds < 2)) exit(-1);
11
12      if (id == 0) {
13          produce(A);
14          flag = 1;
15      }
16      if (id == 1) {
17          while (flag == 0) {
18              // spin through the loop waiting for flag to change
19          }
20          consume(A);
21      }
22  }
```

Figure 11.3: **Pairwise synchronization with incorrect synchronization** – A producer consumer pattern with one thread producing a result that another thread will consume. This program uses a spin-lock to make the consumer wait for the producer to finish. Note: While the logic in this program is correct, it contains a data race. Hence it is not a valid OpenMP program and as written will not work.

of synchronization: *data synchronization* and *thread synchronization*. For data synchronization, we need the two threads to see a consistent value of a variable in memory. In the case of our producer-consumer problem, the variable in question is A. Since a thread can maintain its own temporary view of memory, there is no guarantee that the values created by the producer is written back to memory for the consumer to load. Likewise, there is no guarantee that the consumer will pull a value of A from memory as opposed to picking up a value from inside its own view of the cache hierarchy. Therefore, both threads need to execute events that make their view of shared variables consistent with the view in memory; i.e., both threads need **flush** directives. We need to place a flush directive right after the producer finishes assigning its updated value to A (i.e., after line 13) while the consumer needs to issue a flush right before it consumes the A (i.e., before line 20).

We use full flushes (i.e., without a clause creating a restricted flush set) to restrict how a compiler reorders statements around the flush. The compiler cannot move statements around a flush if the statements involve variables that are part of the flush set. If we did not use a full flush to resolve the data synchronization for A, it would be possible for the compiler to move operations on the flag variable around the producer and consumer functions.

The second aspect of synchronization we need to explore is *thread synchronization* or more precisely how we establish a synchronized-with relationship between the two threads. As we said earlier, we do this with a spin lock. Using the same logic as with our resolution of the data synchronization problem, we need to place flush directives in the right locations to support the spin lock. We show how we might place the flushes to support the spin lock in Figure 11.4. For the producer, we do a full flush after A is assigned. The following statement to set the flag variable is sequenced after the flush of A. That means if another thread does a load of flag from memory as a signal for when it is safe to load A, we can safely assume the flush of A happens-before flag is set.

Inside the consumer thread, we implement the key steps for the spin lock. We have a while loop that spins until the value of flag changes. We need to put a flush before the while loop to make sure we pick up the latest value of flag from memory, but in this case there is no need to do this for anything other than the flag itself. Hence we use the flush with a list defining the restricted flush set. Notice that inside the while loop we need to include an additional flush directive. A compiler does not know anything about multiple threads. It could look at that code and realize that the value of flag is not being written by the iterations of the loop. It is a constant, as far as the compiler is concerned, and it could place it in temporary storage such as a register[1]. Hence, we must place a flush directive inside the loop to force a refresh from memory in order for the spin lock exit event to work.

Once the spin lock resolves, we insert one last flush to make sure we get the latest value of A. The spin lock establishes a synchronized-with relationship. The full flushes for A assures we have correct data synchronization. Therefore, we have a valid happens-before relationship and the program is correct.

Actually, the program **is not** correct. The OpenMP memory model was re-defined to go beyond its original reliance on flushes and barriers. Consistent

---

[1]This decision by the compiler might depend on the compiler optimization level. This could cause a program to "break" just by changing the compiler options.

```
1   int flag = 0;  // a flag to communicate when the consumer can start
2   omp_set_num_threads(2);
3
4   #pragma omp parallel shared(A, flag)
5   {
6       int id = omp_get_thread_num();
7       int nthrds = omp_get_num_threads();
8
9       // we need two or more threads for this program
10      if ((id == 0) && (nthrds < 2)) exit(-1);
11
12      if (id == 0) {
13          produce(A);
14          #pragma omp flush
15          flag = 1;
16          #pragma omp flush (flag)
17      }
18
19      if (id == 1) {
20      #pragma omp flush (flag)
21      while (flag == 0) {
22          #pragma omp flush (flag)
23      }
24      #pragma omp flush
25      consume (A);
26      }
27  }
```

Figure 11.4: **Pairwise synchronization with flushes** – A producer consumer program with a spin lock and explicit flushes. **This code is incorrect** since the operations on the flag define a data race.

with modern programming languages (a topic we will pick up again later in this chapter) a synchronized-with relation cannot be established with regular variables. A synchronized-with relation can only be established with an atomic operation. Setting **flag** to one and later loading it into memory; those are not atomic operation. They involve an unordered read and write to the same address which is a data race and defines a program that is *technically* invalid.

We have tested the program in Figure 11.4 extensively on systems based on the x86 architecture.[2] It works every time. This is because the hardware memory

---

[2]The x86 architecture is an instruction set architecture series for processors developed and used by Intel Corporation.

model supported by Intel is unusually forgiving. It basically treats regular loads and stores as relaxed atomics (that is, atomic operations that are not ordered). Hoping to find cases where the program fails, we moved to Arm CPUs and IBM Power CPUs. These processors have a memory model that is much less forgiving than the x86 memory model and makes no guarantees of atomicity for regular loads and stores. Once again, we found that the program worked every time. Even though the program contains a data race and a program with a data race is invalid, for the case of the program in Figure 11.4 this does not seem to matter. This is a sensitive topic among parallel application developers. According to the designers of programming languages, there is no such thing as a benign data race. If your program has a data race, it is wrong and the compiler does not have to make any guarantees of correctness. For instance, partial differential equation solvers based on relaxation methods sometimes allow data races since synchronization overhead would be too high otherwise[8]. They are iterative and if a data race caused an actual value to be wrong, the error would be corrected in the next few iterations. There are also machine learning algorithms that tolerate data races just to avoid excessive synchronization overhead[12]. These programs work, but that does not matter to people designing languages and building compilers.

How do we make the producer-consumer program correct? The key is to modify the spin lock so the synchronized-with relationship is established through atomic load and store operations. As we discussed in Section 10.2.3, an atomic operation is one that occurs completely or not at all. It is not possible to see an intermediate state for a variable when it is manipulated by atomic operations. In Figure 11.5 we show the changes needed to make the synchronized-with relationship through a spin lock work. For the producer, we put the assignment to the flag variable inside an atomic write construct. We do not need a separate flush for the flag variable following the atomic construct since the implied flush by the atomic construct takes care of making the value of flag consistent with memory. Note, however, that we still need the flush before the atomic. This is needed to assure that the compiler makes the value of A consistent with the value in memory before it carries out the atomic write.

On the consumer side, we had to make a few more changes to support use of the atomic. In Figure 11.5 we turned the while loop into an infinite loop that we break out of once the condition on flag is set. Also to avoid any write conflicts on the variable flag, we store its value into a temporary flag variable and test that variable's value to determine when to break out of the loop. After the spin lock establishes our synchronized-with relation, we do one more flush to be sure we pick up the value of A consistent in memory before the value of flag was set on the producer.

An atomic operation implies a flush so it is debatable if that last flush was needed. The problem is that the OpenMP standard relaxed the rules constraining a compiler's ability to move statements around an implied flush. In essence, the flushes with an atomic do not define an ordering of operations to memory and that extra flush is required. We will revisit this topic later when we talk about the memory model additions to OpenMP inspired by the C++11 standard.

```
1   int flag = 0;  // a flag to communicate when the consumer can start
2   omp_set_num_threads(2);
3
4   #pragma omp parallel shared(A, flag)
5   {
6       int id = omp_get_thread_num();
7       int nthrds = omp_get_num_threads();
8
9       // we need two or more threads for this program
10      if ((id == 0) && (nthrds < 2)) exit(-1);
11
12      if (id == 0) {
13          produce(A);
14          #pragma omp flush
15          #pragma omp atomic write
16              flag = 1;
17      }
18      if (id == 1) {
19          while(1) {
20              #pragma omp atomic read
21                  flag_temp = flag;
22              if (flag_temp != 0) break;
23          }
24          #pragma omp flush
25          consume (A);
26      }
27  }
```

Figure 11.5: **Pairwise synchronization with flushes and atomics** – A producer consumer program with a spin lock and explicit flushes. With the use of atomics to update and then read flag, this program is race free on any processor.

## 11.3   Locks and How to Use Them

A lock in OpenMP has basically the same functionality as a mutex in pthreads. It is used to build synchronization protocols around mutual exclusion. Unlike the `critical` construct, the OpenMP locks are implemented as library routines. This means they can be folded into your software in different ways. We believe one of the easiest cases to appreciate where a lock is needed is when the synchronization protocol is intimately intertwined with a data structure. That is the problem we will address in this section.

First, let's consider the locks themselves. A lock is associated with a variable with the lock type. We show a simple case of using locks in Table 11.3. The lock variable must be initialized before it is used. Once initialized, it can have two values: `set` and `unset`. If a thread calls the routine to set the lock and the lock is already set, it will wait until the lock is unset before setting the lock itself and proceeding. The lock set and unset routines imply a flush so they imply the needed memory movement to support memory consistency with their mutual exclusion functionality.

Table 11.3: **Locks in C/C++ and Fortran.**

| C/C++ | Fortran |
|---|---|
| omp_lock_t lck; | integer (omp_lock_kind) lck |
| omp_init_lock(&lck); | call omp_init_lock(lck) |
| | |
| #pragma parallel shared(lck) | !$omp parallel shared(lck) |
| { | |
|    omp_set_lock(&lck); |    call omp_set_lock(lck) |
|    ... do something |    .... do something |
|    omp_unset_lock(&lck); |    call omp_unset_lock(lck) |
| } | !$omp end parallel |
| omp_destroy_lock(&lck); | call omp_destroy_lock(lck) |

The power of locks emerges when you explore how they interact with different data structures in a program. In Figure 11.6 we show a program that tests a simple uniformly distributed random number generator. It does this by calling the random number generator and constructing a histogram of the returned pseudo-random values. We do not show the logic here, but assume the random number generator is a parallel random number generator that produces a single pseudo-random sequence regardless of how many threads call it.

We create an array, `hist[num_bins]` on line 12. This array holds the bins associated with the histogram. Later in the program, we create a parallel loop (lines 26 to 36) that sets a variable to its next member of the pseudo random sequence and then determines which bin it belongs to in the histogram. The histogram is a shared data structure. Any thread might update any bin in the histogram during any iteration. Using the constructs from the OpenMP Common Core, you would need to put the assignment to the histogram bin inside a critical region which would cause most threads in the team to spend the bulk of their time waiting to access the histogram structure.

The solution is to create an array of locks (line 14) with one lock per bin in the histogram. We initialize the array of locks and the histogram bins in the loop from line 20 to line 23. Essentially, the locks are associated with the histograms, basically becoming part of the histogram itself. If the random number generator is functioning correctly, it will randomly return values across its uniform range. That means the chances of any two threads trying to update the same bin of the histogram at one time is low. In other words, the locks are likely to be uncontended.

In lines 33 to 35, we see how the locks are used. By setting a lock before incrementing a bin, a thread assures that only one thread at a time can execute the update. A flush is implied when a lock is set so the histogram value updated by the thread is consistent with the value in memory. The value of the histogram is flushed when the lock is unset on line 35 so when the next thread grabs the lock to update that bin, it will see the correct value. When we have finished testing the parallel random number, we compute statistics about the distribution of bins in the histogram and destroy the locks in lines 39 to 44.

In the OpenMP specification, there are many variations on locks. This is because locks are the foundational construct for building advanced synchronization protocols. For concurrent algorithm developers, a rich set of lock options is an essential feature of a multithreading programming environment. For parallel application programmers, however, the basic form we have discussed here is usually all that is needed.

```
1   #include <omp.h>
2   #include <math.h>
3   #include "random.h"   \\seed() and drandom()
4   #define num_trials 1000000       // number of x values
5   #define num_bins   100           // number of bins in histogram
6   static long xlow = 0.0;          // low end of x range
7   static long xhi = 100.0;         // High end of x range
8
9   int main ()
10  {
11      double x;
12      long hist[num_bins];   // the histogram
13      double bin_width;        // the width of each bin in the histogram
14      omp_lock_t hist_lcks[num_bins]; // array of locks, one per bucket
15      seed(xlow, xhi);  // seed random generator over range of x
16      bin_width = (xhi − xlow) / (double)num_bins;
17
18      // initialize the histogram and the array of locks
19      #pragma omp parallel for schedule(static)
20      for (int i = 0; i < num_bins; i++) {
21          hist[i] = 0;
22          omp_init_lock(&hist_lcks[i]);
23      }
24      // test uniform pseudorandom sequence by assigning values
25      // to the right histogram bin
26      #pragma omp parallel for schedule(static) private(x)
27          for(int i = 0; i < num_trials; i++) {
28
29          x = drandom();
30          long ival = (long) (x − xlow)/bin_width;
31
32          // protect histogram bins. Low overhead due to uncontended locks
33          omp_set_lock(&hist_lcks[ival]);
34              hist[ival]++;
35          omp_unset_lock(&hist_lcks[ival]);
36      }
37      double sumh = 0.0, sumhsq = 0.0, ave, std_dev;
38      // compute statistics (ave, std_dev) and destroy locks
39      #pragma omp parallel for schedule(static)
40          for (int i = 0; i < num_bins; i++) {
41              sumh   += (double) hist[i];
42              sumhsq += (double) hist[i] * hist[i];
43              omp_destroy_lock(&hist_lcks[i]);
44          }
45      ave = sumh / num_bins;
46      std_dev = sqrt(sumhsq / ((double)num_bins) − ave * ave);
47      return 0;
48  }
```

Figure 11.6: **Locks to protect updates to a histogram** – Generate a sequence of pseudorandom numbers and assigns them to a histogram.

## 11.4   The C++ Memory Model and OpenMP

The C++ memory model defined with the ANSI C++11 standard is the foundational memory model for many modern programming languages. Most programming languages that include threads defined after the ANSI C++11 standard base their own memory models on the C++ standard. OpenMP version 5.0 is no exception to that trend.

The problem is that the C++ memory model is exceedingly confusing. Very few people, even self-described experts, understand the model in full. It is powerful and lets programmers write complex concurrent algorithms with low overheads. Using its more subtle features correctly, however, without introducing race conditions can be very difficult.

We will not attempt a full description of the ANSI C++11 memory model. Instead we will provide an overview of the features of the model that appear in the OpenMP version 5.0 standard. The central idea in understanding how this model projects into OpenMP is one we have discussed already. A memory model is defined in terms of happens-before relations, data synchronization operations, and synchronized-with operations based on the behavior of atomics. This is a major departure from the original OpenMP memory model defined in terms of flush operations (both implicit and explicit).

The newer model is complicated and way beyond the ability of general parallel application developers to understand. It would have been nice if we stayed with the simpler models OpenMP started with. The view inside the OpenMP language communities, however, was that models based on flushes incur too much overhead. While easier for programmers to understand, the performance impact was intolerable and a more fine-grained model based on the behavior of atomics and flushes was needed.

C++11 (and versions of the language since then) defined atomic operations that are used to define synchronized-with relations. As we stated earlier, an atomic operation is a special type of operation that either runs to completion or does not happen at all. In other words, a program can never observe a partially complete atomic operation. Examples of atomic operations include load (read), store (write), fetch and add, exchange, and fetch with the familiar logical operations AND, OR, and XOR.

To play their role as the foundation of synchronization in a programming language, the observable orders of atomic operations with respect to the order of loads and stores of other variables must be defined. The most commonly used memory orders in C++ include the following:

- seq_cst or sequentially consistent: Loads and stores to memory will appear to all threads to occur in the same order. This order will be any semantically valid interleaving of instructions executed on all of the threads.

- release: store operations (both atomic and non-atomic) that are sequenced before a release operation, R, may not be reordered to appear to occur after R.

- acquire: load operations (both atomic and non-atomic) that are sequenced after an acquire operation, A, may not be reordered such that they appear to occur before A.

- acquire_release: loads and stores (both atomic and non-atomic) may not be reordered around an acquire_release operation.

These memory orders are indicated as an additional parameters to the atomic construct as shown in Table 11.4.

Table 11.4: **Atomic constructs in C/C++ and Fortran** – These are the same as the atomic constructs defined earlier but with an optional clause to define the memory order.

| |
|---|
| **#pragma omp atomic** *[atomic_clause] [memory_order_clause]* |
| **!$omp atomic***[atomic_clause] [memory_order_clause]* |
| where *atomic_clause* is read, write, update, or capture |
| and *memory_order_clause* is seq_cst, acq_rel, release, acquire, or relaxed |

If an atomic operation occurs without any of the above memory order clauses, it is said to be a relaxed atomic. That means the operation is still atomic (i.e., it either executes to completion or not at all), but it does not imply an order with respect to loads and stores to memory. Regardless of the memory order defined with an atomic operation, atomic operations on the same object may never be reordered with respect to each other.

A complete discussion of the memory orders and how to use them would fill a book. To understand them, we recommend the classic book *C++ Concurrency in Action* [15]. In this excellent (but dense) book, the memory orders from the C++11 memory model are described as are their use in concurrent algorithms and data structures. Of particular interest is the fascinating problem of how to build concurrent data structures that support proper data synchronization but without the use of locks. While fascinating, this type of programming is extremely advanced and best left to people who make a full-time job of writing such code. In other words, NOT the application programmers OpenMP was created to serve.

While it is good to be aware of all the memory orders from the C++ programming language, we recommend only using one of them: the sequentially consistent memory order. This is the intuitive memory order most people working with multithreaded programming assume to be the case. You have a number of concurrent threads. Since they are concurrent, the operations between threads are unordered with respect to each other outside of explicit thread synchronization operations (such as a barrier or a spin-lock). The operations, as observed through updates of shared variables in memory, appear to carry out those updates as if the instructions across threads are interleaved in some semantically allowed order. As long as the program is race free, this simple memory order is easy to reason about and use.

Let us return to the producer-consumer program with spin-lock to establish a synchronized-with relation. In Figure 11.7, the atomic operations on lines 14 and 20 look the same as before but we have added `seq_cst`, a memory order clause. This states that the memory order of loads and stores around the atomics will follow that implied by a sequentially consistent program. That means updates to shared variables cannot be moved by the compiler around the atomic operations. Instructions sequenced-before the atomic operations complete before the atomic operation. Sequence points that follow the atomic operation do not occur until after the atomic operation. Memory movement operations needed to support the interleaved semantics are implied; therefore explicit flushes are not required.

The `seq_cst` memory order was added to OpenMP in version 4.5. It is supported in most if not all OpenMP compilers. The other memory orders were not added until OpenMP 5.0 and therefore require compilers that support OpenMP 5.0 or later. There can be efficiency benefits to using the other memory orders. An atomic with release semantic pairs with an atomic with acquire semantics to define a synchronized-with relation. This is expected to add significant efficiency benefits relative to the `seq_cst` memory order.

We do not recommend, however, using the other memory orders at this time. They are difficult to understand. Errors arising from problems with order of operations

```
1    int flag = 0;   // a flag to communicate when the consumer can start
2    omp_set_num_threads(2);
3
4    #pragma omp parallel shared(A, flag)
5    {
6        int id = omp_get_thread_num();
7        int nthrds = omp_get_num_threads();
8
9        // we need two or more threads for this program
10       if ((id == 0) && (nthrds < 2)) exit(-1);
11
12       if (id == 0) {
13           produce(A);
14           #pragma omp atomic write seq_cst
15               flag = 1;
16       }
17
18       if (id == 1) {
19           while(1) {
20               #pragma omp atomic read seq_cst
21                   flag_temp = flag;
22               if(flag_temp != 0) break;
23           }
24           consume(A);
25       }
26   }
```

Figure 11.7: **Pairwise synchronization with sequentially consistent atomics**
– A producer consumer program but now the form of atomic construct used implies all the
flushes we need.

in memory are notoriously difficult to debug. The bugs produced are sporadic and
difficult to expose in all but the most rigorous testing regimes. We recommend
sticking with the more straightforward seq_cst and only turn to the other memory
orders if scalability problems in your algorithm demand that you do so. Note that in
most scalable algorithms, the work between synchronization operations dominates
the run time and small performance differences in synchronization protocols do not
weigh heavily on the program's overall performance.

## 11.5   Closing Comments

Experienced multithreaded programmers typically do not understand the low level details of memory models. Most of us use a simplified set of memory model rules and rarely descend into the mysterious depths of a memory model. We strongly urge you to do the same.

To drive this point home, we suggest the following guidelines:

- As your first course of action, use collective synchronization operations such as critical sections and barriers. Be very cautious with the `nowait` clause on worksharing constructs[3].

- If pairwise protocols are required, use atomics with the sequentially consistent memory order. Avoid explicit flushes.

- Understand the distinction between data synchronization and thread synchronization. Barriers (both implicit and explicit) combine the two making their use particularly simple. Mutual exclusion constructs (critical sections and locks) are best used only for data synchronization. For thread synchronization, use atomics with the sequential consistent memory order.

The last point deserves further explanation. In OpenMP 3.1, the rules constraining how a compiler can reorder instructions around an implicit flush were relaxed. This was done because the more severe restrictions inhibited the ability of a compiler to extract sufficient performance from a program. Therefore, other than for the barrier, a programmer must consider the flushes implied by certain OpenMP constructs to support the data synchronization associated with the construct. Do not use anything other than a barrier or an atomic with a sequentially consistent memory order to establish a synchronized-with relation for thread-synchronization.

---

[3] A `nowait` disables the implicit barrier at the end of a worksharing construct. By disabling the barrier, the `nowait` would also eliminate the flush implied by the end of a worksharing construct which could break some programs.

# 12 Beyond OpenMP Common Core Hardware

When we created OpenMP, shared memory systems were typically viewed as symmetric multiprocessors (SMP). The operating system of an SMP treats each processor equally and assumes that the cost of getting to any location in memory is the same across the system. We knew this was a crude approximation. A memory hierarchy populated with multiple levels of cache was the norm even in the late 1990s when we started work on OpenMP. Experience at that time had convinced us, however, that taking the non-uniform cost for memory accesses into account was too much for most programmers to handle, so we stuck with the SMP model.

Current platforms are built from processors with dozens of cores and multiple processors sharing a memory hierarchy on a single node. Programmers can no longer pretend their systems follow the SMP model. If achieving a large fraction of the performance available from a system is the goal, programmers have no choice but to write code that takes the non-uniform memory access (NUMA) features of computers into account.

Modern hardware design is challenging programmers in ways that go well beyond the need to design code around the NUMA features of a system. Hardware trends emphasize parallelism beyond basic multithreading. For example, the silicon area dedicated to vector units on a CPU has grown to the point where programmers can no longer afford to ignore them. Compilers have been automatically generating vector code for decades, but even after years of research on the problem, compiler-driven autovectorization does not effectively exploit the vector units for most applications. Programmers need a way to tell a compiler how to vectorize code. We do this in OpenMP by thinking about the vectorization problem in terms of the Single Instruction Multiple Data (SIMD) model. A number of directives and supporting clauses have been added to OpenMP to cover these cases.

Finally, the hardware landscape faced by programmers has changed radically as the GPU has moved from a specialized graphics processor to a powerful general purpose processor for data parallel algorithms. Programmers need to write software for GPUs and OpenMP has grown to cover these devices.

In this chapter, we explore these three branches of modern hardware development and describe how OpenMP addresses them. We will not cover them in enough detail for you to master programming these systems. Each of these topics deserves several chapters if not a whole book! Our goal in this chapter is to provide an overview, so you can understand the implications of using OpenMP on these systems.

## 12.1   Nonuniform Memory Access (NUMA) Systems

In Chapter 1, we discussed the fundamental hardware model we had in mind as we first created OpenMP. We called this a symmetric multiprocessor (SMP). There are two features that make a computer an SMP. First, the time to access a variable from any processor to any memory location must be the same. We call this a Uniform Memory Access (UMA) system. Second, the operating system must treat all processors equally.

As we explained in Chapter 1, the SMP model is an oversimplification. A modern processor is not an SMP system. Locations in memory do not have an equal cost from each core to any address in memory due to the cache hierarchy.

It is more accurate to consider modern processors as Non-Uniform Memory Access (NUMA) systems. For a NUMA system, the cost of accessing a given address in memory varies from one processor (or core) to another. An individual CPU is a NUMA system, but when we put two or more CPUs together into a single node, the NUMA features of the system become quite complicated.

We show a block diagram of a typical server-node in Figure 12.1. This server-node has two sockets each of which holds an Intel® Xeon™ E5-2698 v3 CPU. Each CPU has 16 cores with two hardware threads per core. A hardware thread is a set of architectural features (including registers and buffers) needed to support the state of a thread. This means that a single core can maintain two active threads at the same time, an approach known as simultaneous multithreading (SMT). There is still a single Arithmetic Logic Unit, but the core supports two active threads so if one thread is blocked, the core can extract useful work (and keep the core's resources busy) from the other thread. To the operating system, each hardware thread appears as a distinct place to execute a thread; therefore, each core is counted as a pair of logical CPUs by the OS.

The cores are organized around an on-chip network implemented as a pair of rings. Each core has an L2 unified cache and a pair of L1 caches (one for data and one for instructions). The Xeon™ E5-2698 CPU has a shared L3 cache implemented as a large number of blocks, one block per core. Hence, the NUMA features of the CPU even impact the L3 cache since a cache line in L3 is quicker to access if it comes from the block associated with a core as opposed to a block on a core many hops across the on-chip network.

For our discussion of NUMA systems, however, the most important feature of a server-node is the memory controllers. Each CPU has a pair of memory controllers

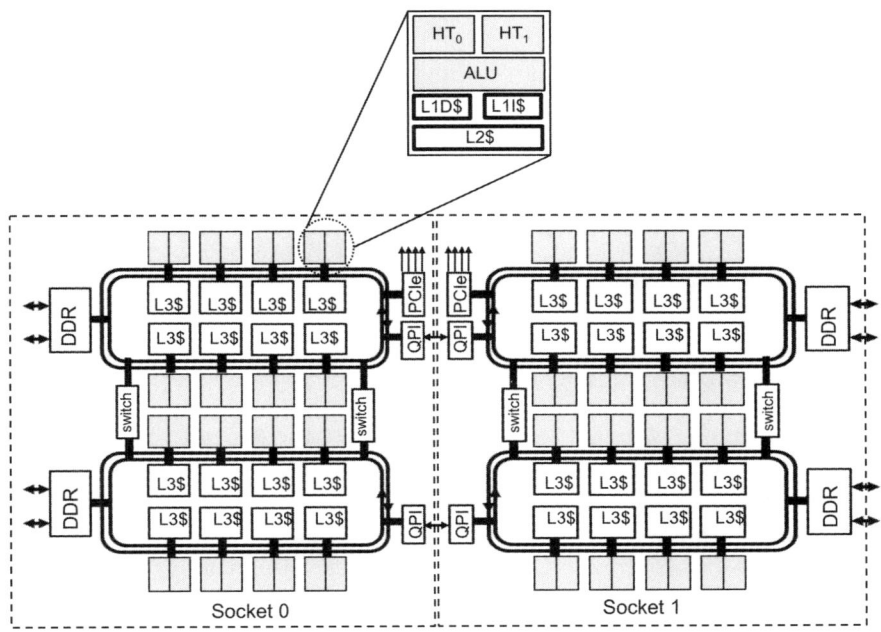

Figure 12.1: **A typical NUMA server-node** – The server-node used in the examples in this chapter with two 16-core Intel® Xeon™ E5-2698 v3 CPUs (Haswell) at 2.3 GHz with 128 GB DDR4 2133 MHz memory (four 16 GB DIMMs per socket). There are 2 NUMA domains per server-node with 16 cores per NUMA domain. We show an expanded view of one of the cores for the details of how 2 hardware threads per core share the unified L2 cache and the L1 data cache and the L1 instruction cache.

with their own blocks of dynamic random access memory (DRAM). For the server-node in Figure 12.1, the DRAM is based on DDR memory (Double Data Rate synchronous DRAM). A CPU sits in a socket. The two CPUs are connected by a high speed interconnect (QPI or "Quick Path Interconnect") to maintain a single address space between the CPUs. Memory access times, however, are much less when a core accesses a variable in the memory attached to its memory controller as opposed to a variable in memory associated with the CPU in the other socket. We say that the server-node has two NUMA domains, one associated with each socket.

This discussion has covered a level of detail, well beyond what most programmers would ever care to know. We wanted to show these details, however, to drive home a key point. NUMA effects are complicated and show up at many levels: the L3 cache blocks, the four memory controllers on the server-node, and even the L1 caches on a single core. Given this complexity, it is amazing that programming to the SMP model works at all. It does work, however, since the operating system working closely with the OpenMP runtime does an excellent job of masking these effects. There are times, however, when no amount of runtime system cleverness can hide the NUMA nature of a system. To get a high fraction of peak performance, you have to organize the threads and the data structures in your program around the NUMA features of a system.

### 12.1.1   Working with NUMA Systems

A thread runs on a hardware resource. We often use the name *processor* to refer to a generic hardware resource in a computer system. For this discussion, however, we want to be very careful with our terminology and avoid confusion that arises from mixing the terms cores, processors, CPUs and hardware threads. Therefore, we will define a new and more precise term, *place*, to refer specifically to any hardware resource that supports the execution of a thread.

In an SMP system, there is little reason to think about the places where threads run. As long as they are evenly spread out and stay out of each other's way, where threads run is not important. For a NUMA system, however, it is critical to understand the places threads run. We use the term *thread affinity* to refer to mapping threads onto particular places and binding them to those places so the OS cannot migrate the threads to different places.

To manage thread affinity, we need to understand how to refer to the hardware resources in our system. OpenMP threads are managed on behalf of our program by the operating system. The operating system views a NUMA server node as a collection of logical CPUs; that is, each hardware thread is a logical CPU to the OS. If we are going to manage thread affinity, we need to identify specific hardware threads by assigning numbers to them.

There are a number of tools that help us understand the places in a node. The Linux tool, *numactl* is used to control the NUMA policy for processes and shared memory. To find information about the NUMA characteristics of a processor on a node, use the command:

```
$ numactl -H
```

Figure 12.2 shows the output from "numactl -H" on the server-node in Figure 12.1. It reports that each server-node is composed of 2 "nodes" referring to two NUMA domains. These are numbered from 0 to 1. NUMA node 0 has logical CPUs 0-15 and 32-47. NUMA node 1 has logical CPUs 16-31 and 48-63. The tool also reports the total and available memory sizes on each NUMA domain and the relative distances from CPUs on each NUMA domain to access memory in both NUMA domains. It is clear that memory distance from NUMA domain 0 to 0 is closer than from NUMA domain 0 to 1. Note that the values 10 and 21 are relative numbers; they do not necessarily mean that the ratio of the distances is exactly 21 to 10.

```
1   % numactl -H
2   available: 2 nodes (0-1)
3   node 0 cpus: 0 1 2 3 4 5 6 7 8 9 10 11 12 13 14 15 32 33 34 35 36
4   37 38 39 40 41 42 43 44 45 46 47
5   node 0 size: 64430 MB
6   node 0 free: 63002 MB
7   node 1 cpus: 16 17 18 19 20 21 22 23 24 25 26 27 28 29 30 31 48
8   49 50 51 52 53 54 55 56 57 58 59 60 61 62 63
9   node 1 size: 64635 MB
10  node 1 free: 63395 MB
11  node distances:node   0   1
12  0:  10  21
13  1:  21  10
```

Figure 12.2: **"numactl -H" result for an Intel® Xeon™ CPU E5-2698 v3 (Haswell) at 2.3 GHz server node**.

A second tool often used when working with NUMA systems is the portable hardware locality (hwloc) tool from the Open MPI project. This tool provides information about the system topology (i.e., how different nodes in a cluster are connected), the features of the NUMA nodes in the system, cache information, and the mapping of processors onto nodes. The command:

```
$ hwloc-ls
```

provides a textual representation of the output from hwloc. You can use the command:

```
$ lstopo
```

to see a graphical representation of the hwloc tool data. We show this data in Figure 12.3 for the server-node we have been discussing (Figure 12.1). The image shows that each node has 2 NUMA domains (1 NUMA domain per socket). There are 16 physical cores per NUMA domain, and 2 hardware threads per core. Each core has an L2 cache (256 KB) and two L1 caches (32 KB instruction cache and 32 KB data cache). There is a 40 MB shared L3 cache per socket.

The core numbering scheme in this figure represents the logical CPU numbers. For example, NUMA domain 0 has 16 cores: physical cores 0 to 15. This particular processor supports two hardware threads per core giving us 32 logical cores. How numbers map onto cores is implementation dependent, hence why we need tools to discover this mapping. For the server node we have been working with (Figure 12.1), the numbers are assigned successfully to each physical core and then wraps around. Therefore logical CPUs 0 and 32 are on physical core 0, logical CPUs 1 and 33 are on physical core 1, and so on. This same scheme is used for NUMA domain 1 which also has 16 cores: physical cores 16 to 31. Logical CPUs 16 and 48 are on physical core 16, logical CPUs 17 and 49 are on physical core 17, and so on. We show the full numbering for the logical CPUs in Figure 12.4. As we will see later in this chapter, this information on the numbering of logical CPUs will be critical for managing thread affinity.

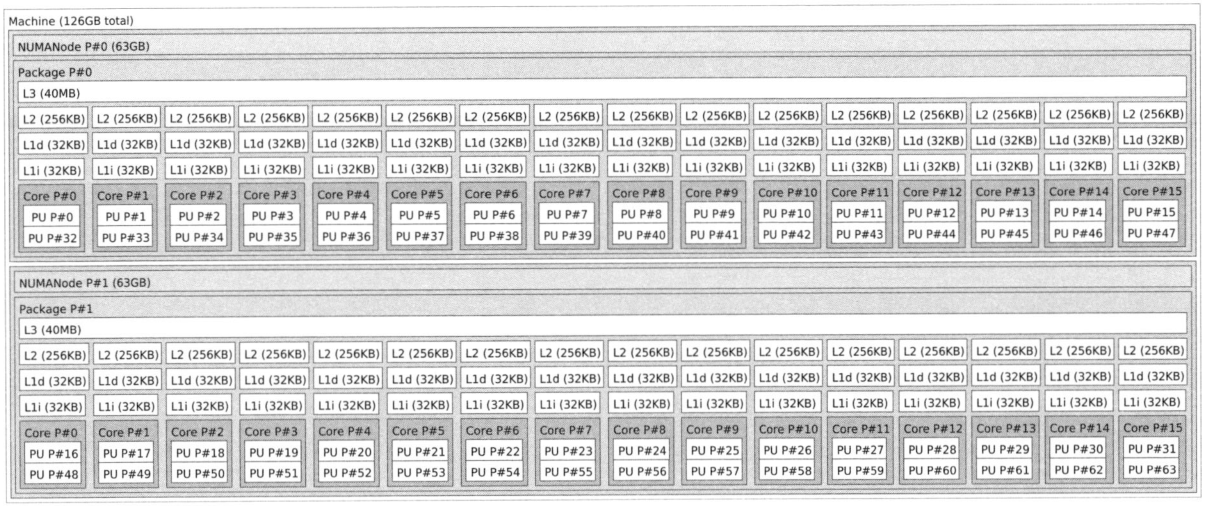

Figure 12.3: **Graphic representation of a 32-coreIntel® Xeon™ Haswell CPU E5-2698 v3 at 2.3 GHz compute node from "lstopo"** – It shows that each node has 2 sockets (1 NUMA domain per socket). There are 16 physical cores per NUMA domain, and 2 hardware threads per core. Each core has a unified L2 cache and a pair of L1 caches (32 KB instruction cache, 32 KB data cache). There is also a shared 40 MB L3 cache per socket.

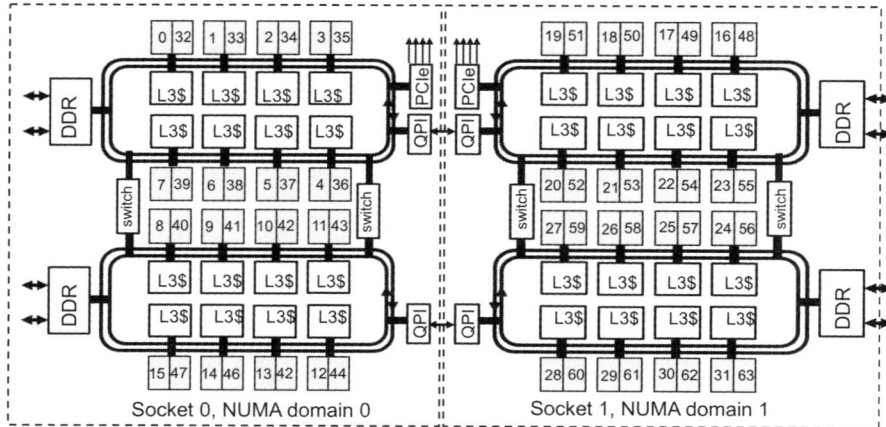

Figure 12.4: **A typical NUMA server-node with logical CPU numbers** –
A server with two 16-coreIntel® Xeon™ E5-2698 v3 CPUs showing how the operating
system maps logical CPU numbers onto hardware threads.

#### 12.1.1.1 Controlling Thread Affinity

In OpenMP you control thread affinity through two closely related concepts: *places*
and *processor-binding*. We have already explained the concept of places. A place is
a hardware resource upon which a thread can execute. Examples include a socket, a
core, or a hardware thread. Processor-binding is a policy that describes how threads
are (or are not) mapped onto places and whether they are allowed to migrate during
the execution of a program.

We will start by explaining how places work in OpenMP. You communicate a set
of places primarily through an environment variable OMP_PLACES. You can define
places explicitly by assigning lists of logical CPU IDs to OMP_PLACES. For example,
you can define four groups of places with specific IDs:

```
export OMP_PLACES="{0,1,2,3},{4,5,6,7},{8,9,10,11},{12,13,14,15}"
```

It can be tedious to list each and every ID. Ranges can be specified using the format
{lower-bound:length:stride} with a default stride of one. Using this approach
we can define the same set of places as above with the notation:

```
export OMP_PLACES="{0:4},{4:4},{8:4},{12:4}"
```

To understand strides other than 1, we'll consider one more example. We can represent the place-list {0,2,4} as {0:3:2}.

We usually do not recommend the use of explicit lists. Instead, you can define a category of places and let the OpenMP runtime do the actual assignments as it carries out processor binding. In these cases, you would set the environment variable OMP_PLACES to one of the values:

- **threads**: Bind OpenMP threads at the granularity of hardware threads.

- **cores**: Bind OpenMP threads at the granularity of cores.

- **sockets**: Bind OpenMP threads at the granularity of sockets.

To understand how the above OMP_PLACES values result in the placement of threads, we have to explain processor binding. You can control processor binding through an environment variable OMP_PROC_BIND. The values of this environment variable set the affinity policy and define how threads should be scheduled onto places.

- **true**: Thread affinity is enabled with an implementation defined default place list.

- **false**: Thread affinity is disabled.

- **master**: Each thread in the team is assigned to the same place as the master thread of the team.

- **close**: The threads in the team are placed close to the master thread. Threads are assigned to consecutive places in a round-robin fashion starting with the place to the right of the master thread.

- **export OMP_PROC_BIND=spread**: Spread the threads as evenly as possible over the places.

Once the policy is set and the thread binding is determined, threads are not allowed to migrate outside the places to which they are assigned. We show some examples to help clarify how thread affinity works in OpenMP. In Figure 12.5 we show how threads bind to CPUs using a CPU with 4 cores, 2 hardware threads per core, and OMP_NUM_THREADS set to 4.

Controlling thread affinity through environment variables impacts the entire program. You can override these "whole program" controls using the proc_bind

## Close: Bind threads as close to each other as possible

| Node | Core 0 | | Core 1 | | Core 2 | | Core 3 | |
|------|-----|-----|-----|-----|-----|-----|-----|-----|
|      | HT1 | HT2 | HT1 | HT2 | HT1 | HT2 | HT1 | HT2 |
| Thread | 0 | 1 | 2 | 3 | | | | |

## Spread: Bind threads as far apart as possible

| Node | Core 0 | | Core 1 | | Core 2 | | Core 3 | |
|------|-----|-----|-----|-----|-----|-----|-----|-----|
|      | HT1 | HT2 | HT1 | HT2 | HT1 | HT2 | HT1 | HT2 |
| Thread | 0 | | 1 | | 2 | | 3 | |

Figure 12.5: **Binding 4 OpenMP threads onto a CPU** – The CPU has 4 cores with 2 hardware threads per core. With OMP_PROC_BIND set to "close", 4 threads bind to the first 2 physics cores, using both hardware threads per core. With OMP_PROC_BIND set to "spread", 4 threads bind to all 4 physical cores, using only 1 hardware thread per core.

clause on a parallel construct. The clause takes the same `master`, `spread`, or `close` values as used with the OMP_PROC_BIND environment variable. The following is an example of how `proc_bind` and the `num_threads` clauses are used together on a `parallel` construct to manage thread affinity.

```
C/C++:
    #pragma omp parallel num_threads(2) proc_bind(spread)
```

```
Fortran:
    !$omp parallel num_threads(2) proc_bind(spread)
        ...
    !$omp end parallel
```

### 12.1.1.2   Managing Data Locality

We have solved half of the NUMA problem: we know how to control where OpenMP threads execute in our NUMA system. To complete our NUMA programming story, we need to understand how data maps onto those same hardware resources. This is the issue of *data locality*.

The general idea is to try to keep data local to the places where the work on that data will occur. Data locality is composed of two components: *cache locality* and *memory locality*. We have discussed cache locality before. Any time you bring data into a cache, you need to structure your computations so they reuse data in cache. The new idea for our NUMA discussion is memory locality.

Memory locality is the degree to which data resides in a block of DRAM that is close to the processors working with that data. To describe how memory locality works, we need to explain a few concepts about the organization of memory in a computer.

CPUs used in modern servers use 48-bit addresses. In principle, you can reference over 268 TeraBytes of memory with 48-bit addresses. This is way beyond even the largest available DRAM memories. A computer system, therefore, defines a virtual memory system organized around contiguous blocks of memory called pages. Pages range in size from 4 KiloBytes (a common default size) to MegaBytes or even in rare cases, up to GigaBytes. Choosing the right page size is complicated and raises issues well beyond the scope of this discussion. The point to remember is that a page easily fits in DRAM and is swapped between DRAM and the secondary storage system (a spinning disk or an SSD solid state storage device) so the memory addressed by an OS exceeds the total size of DRAM.

When you work with data in memory, you are working within blocks of memory organized into pages. A page is swapped into active use and is allocated to a range of physical addresses in memory (i.e., in DRAM). There are a number of policies that control this association between a page and its location. The most common policy, which is the default on most Linux systems, is called *first touch*. Using the terminology we have established in this chapter, the first touch policy says that a page of memory is associated with the memory controller closest to the thread that first accesses the memory. In other words, when a thread first touches a location in memory, the page that holds that location is moved to the physical memory close to that thread.

For a programmer, the way you work with a system using the first touch policy is straightforward. You first touch data when you initialize it. In other words, it is *not* the allocation of the block of memory (using, for example, a `malloc()` function) that determines where a page will be located. The key is how that data is first initialized. Therefore, to make sure the pages in the virtual memory system are located in physical memory (i.e., DRAM) where you need them, you must initialize your data with the same threads that will later work with that data.

To explore these issues we will use the triad kernel from the STREAM benchmark [10]. This uses a simple operation where a vector is scaled by a value and the result is added to another vector:

```
for (j = 0; j < VectorSize; j++) a[j] = b[j] + d * c[j];
```

The floating point operations (one add and one multiply) take very little time compared to the time required to move data from memory. The performance is bounded by the memory bandwidth, which is indeed what the STREAM benchmark is designed to expose. We will run this benchmark after initializing the data in one of two ways. In the first, the data is touched first by the initial thread before the `parallel` construct creates the team of threads that will work with the data. This is Step `1.a` in Figure 12.6. In the second (Step `1.b` in Figure 12.6), we initialize the data with the same set of threads and with the same schedule of loop iterations that we will later use for the triad operation. This assures that we are using the first touch policy to keep data close to the threads that will use it. Notice that we had to disable dynamic mode to make sure the system would use the same threads between the parallel regions. We also needed to use a `static` schedule on the loop since OpenMP guarantees that if the size of the team of threads is the same between parallel for regions and the loop schedule is static, the mapping between loop iterations and the threads will be the same.

Figure 12.7 shows the results of running the STREAM triad benchmark with and without first touch on the server-node described in Figures 12.1 and 12.3 (two NUMA domains each with an Intel® Xeon™ E5-2698 v3 CPU with 16 cores and two hardware threads for a total of 32 logical cores per NUMA domain). The rules for the STREAM benchmark stipulate that the size of the arrays used for the benchmark must be at least 1 million Bytes or 4 times the size of the shared last level cache (whichever of these is larger). For these experiments, the processors had a shared 40 MB L3 cache, so we used an array size (i.e., VectorSize) of 64,000,000 which exceeded the limits required by the benchmark.

When the number of threads is smaller than or equal to the number of logical cores in a NUMA domain, which is 32 in this case, with `OMP_PROC_BIND=close` and `OMP_PLACES=threads`, all the threads will bind to the first NUMA domain, the STREAM bandwidth with and without first touch are the same. However, when the number of threads is larger than 32, cores from the second NUMA domain will be

```
1   //Step 1.a Initialization by initial thread only
2       for (j = 0; j < VectorSize; j++) {
3           a[j] = 1.0; b[j] = 2.0; c[j] = 0.0;}
4
5   //Step 1.b Initialization by all threads (first touch)
6       omp_set_dynamic(0);
7       #pragma omp parallel for schedule(static)
8       for (j = 0; j < VectorSize; j++) {
9           a[j] = 1.0; b[j] = 2.0; c[j] = 0.0;}
10
11  //Step 2 Compute
12      #pragma omp parallel for schedule(static)
13      for (j = 0; j < VectorSize; j++) {
14          a[j] = b[j] + d * c[j];}
```

Figure 12.6: **STREAM initialization with and without first touch** – Without first touch: step 1.a + step 2. With first touch: step 1.b + step 2.

used. The results from the first touch case continue to improve STREAM bandwidth performance since memory access is still going to a local NUMA domain for each thread. Without first touch, however, the bandwidth saturates as threads are forced to cross the boundaries between NUMA domains and access pages mapped to a non-local range of physical memory.

In Figure 12.8 we examine the effect of OMP_PROC_BIND for the STREAM benchmark, once again on the Intel Xeon CPU based server-node described in Figures 12.1 and 12.3 with the code from Figure 12.6. We set OMP_PLACES=threads for all cases. The STREAM bandwidth performance with OMP_PROC_BIND=spread outperforms the performance with OMP_PROC_BIND=close for any number of threads from 2 to 63. When the number of threads is equal to 1 or the number of logical CPUs (64), there is no difference between the two cases and the performance (as we observe) should be the same.

We consider two cases: (1) the number of OpenMP threads is 32 or less, and (2) the number of OpenMP threads is greater than 32. The number "32" is special in this case since that is the number of hardware threads (i.e., logical CPUs) in a NUMA domain for this server-node.

For the first case, that is, when the number of OpenMP threads are 32 or less, when OMP_PROC_BIND=close and OMP_PLACES=threads, OpenMP threads are added to adjacent hardware threads. In other words, the hardware threads on a core are

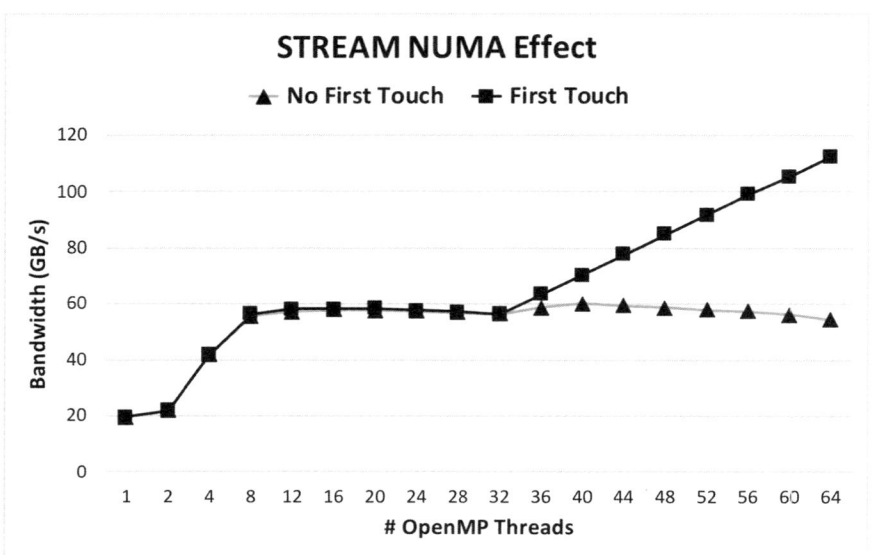

Figure 12.7: **STREAM bandwidth performance with and without first touch** – Results are shown for the node in Figure 12.1 which has 32 cores per node and 2 NUMA domains with 16 cores per NUMA domain and 2 hardware threads per core. Using the code from Figure 12.6, we run STREAM triad without first touch (step 1.a + step 2) and with first touch (step 1.b + step 2). We used OMP_PLACES=threads and OMP_PROC_BIND=close for both cases. The first touch implementation continues to improve STREAM bandwidth performance when the number of threads is beyond 32 (the number of logical cores in a single NUMA domain).

used before going to the next core. The OpenMP threads remain in a single NUMA domain and once both memory controllers in that NUMA domain are engaged, the bandwidth saturates and does not improve until we go beyond 32 OpenMP threads. For OMP_PROC_BIND=spread, threads bind only to the first hardware thread on any core. This spreads out the OpenMP threads among the cores. We see a big jump from 4 to 8 OpenMP threads since at that point, we exploit the second memory controller in the NUMA domain. From 16 to 31 OpenMP threads for OMP_PROC_BIND=spread, threads are added to the second NUMA domain. This means they are accessing the additional memory controllers in the second NUMA domain and we see roughly double the bandwidth compared to the case of OMP_PROC_BIND=close.

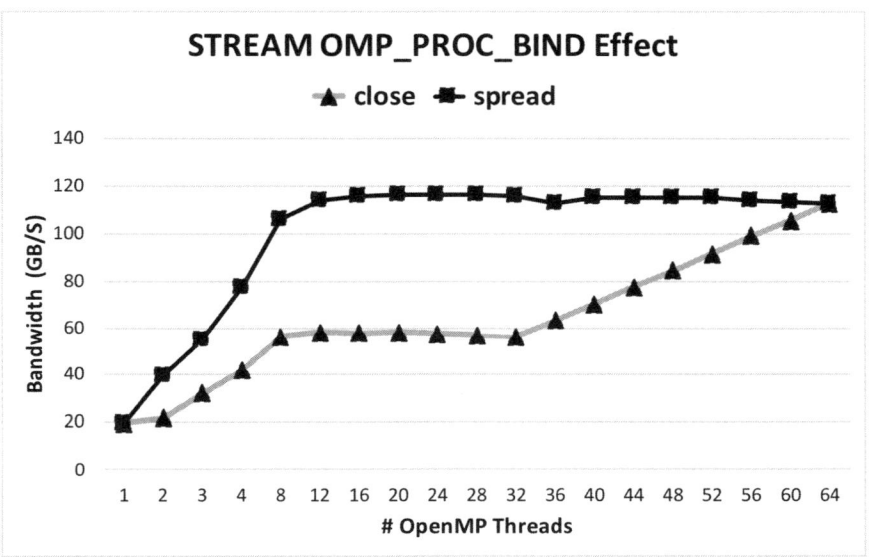

Figure 12.8: **STREAM bandwidth performance with OMP_PROC_BIND settings of close and spread** – Results are shown for the node in Figure 12.1 which has 32 cores per node and 2 NUMA domains with 16 cores per NUMA domain and 2 hardware threads per core. We use First Touch and `OMP_PLACES=threads` for both `OMP_PROC_BIND` settings. The STREAM bandwidth performance with `OMP_PROC_BIND=spread` outperforms the performance with `OMP_PROC_BIND=close` for any number of threads from 2 to 63. Both settings have the same performance when the number of threads are 1 and 64.

When the number of threads is larger than 32 with `OMP_PROC_BIND=close`, the threads continue to be added one hardware thread at a time, but now they move to the second NUMA domain and benefit from the pair of memory controllers on that socket. As seen in Figure 12.8, the STREAM bandwidth performance steadily improves and at 64 OpenMP threads, matches that found for `OMP_PROC_BIND=spread` since at that point, the two cases are identical. For `OMP_PROC_BIND=spread`, the performance does not improve beyond 32 threads since by that point, all the memory controllers in both NUMA domains were already fully engaged; there was no additional bandwidth to add as threads were added since the bandwidth was already saturated.

Selecting the right combination of first-touch, processor binding, and the granularity for adding threads is challenging. Spreading threads out helps exploit available memory bandwidth. However, it can increase synchronization overhead between threads that are further apart. Putting threads near each other reduces synchronization overhead and improves cache reuse, but at the cost of decreased memory bandwidth. To find the best policies, you need to experiment and try different options until you find the optimal combination.

### 12.1.2   Nested Parallel Constructs

You can also influence the distribution of threads in a NUMA system by using nested parallel constructs. In Figure 12.9 we provide an example of an OpenMP program that contains nested parallel regions. In this example, there are 3 levels of nested OpenMP parallel regions, 2 threads in each level. The num_threads clause is used to specify the number of threads desired for each parallel region.

We ran the nested parallelism program from Figure 12.9 in two different ways. The results are shown in Figure 12.10. For our first run, the first parallel region created two threads, but all the nested parallel regions ran with only one thread. On the system where we ran these tests, nested parallelism with OpenMP was disabled. The OpenMP specification leaves it up to the system to decide if nested parallelism is enabled or not by default. To run our nested parallelism program and actually create nested teams of threads, we must tell the system to enable nested parallelism. It is also advised (though not always necessary) to think about the needs of your algorithms and specify the number of levels of nesting as well. Therefore, for the second run of our nested parallelism program, we set the following pair of environment variables: OMP_NESTED [1] and OMP_MAX_ACTIVE_LEVELS.

```
export OMP_NESTED=true
export OMP_MAX_ACTIVE_LEVELS=3
```

We now see the desired result in the second part of Figure 12.10. We create two threads in the first level. Each of those threads creates two threads in the second level, each of which then creates two threads in the third level.

---

[1] Notice OMP_NESTED is deprecated in OpenMP 5.0. Nested mode is automatically enabled when you specify OMP_MAX_ACTIVE_LEVELS value larger than 1 or when OMP_NUM_THREADS or OMP_PROC_BIND indicates nested parallelism.

```
1  #include <omp.h>
2  #include <stdio.h>
3  void report_num_threads(int level)
4  {
5     #pragma omp single
6     {
7         printf("Level %d: number of threads in the team: %d\n", \
8                 level, omp_get_num_threads());
9     }
10 }
11 int main()
12 {
13    omp_set_dynamic(0);
14    #pragma omp parallel num_threads(2)
15    {
16        report_num_threads(1);
17        #pragma omp parallel num_threads(2)
18        {
19            report_num_threads(2);
20            #pragma omp parallel num_threads(2)
21            {
22                report_num_threads(3);
23            }
24        }
25    }
26    return(0);
27 }
```

Figure 12.9: **Nested OpenMP parallel constructs** – There are 3 levels of nested OpenMP parallel regions, 2 threads in each level. The num_threads clause is used to specify the number of threads desired for each parallel region.

The OMP_NUM_THREADS, OMP_PLACES, and OMP_PROC_BIND environment variables were extended to support nesting. OMP_NUM_THREADS can be set to a list of values for successive nesting levels. For example:

```
export OMP_NUM_THREADS=8,4,2
```

The internal control variable for the number of threads (*nthreads-var*) is set to the first number in the list (8 in this example). After the parallel region is created using that value of *nthreads-var*, it is reset to the next value in the list (4 in this case). After the next parallel region is created, *nthreads-var* is set to the next value in the list (2 in this case). This continues until we run out of values in the list at which point the *nthreads-var* ICV remains set to the last item in the list.

```
 1  % ./nested-omp
 2  Level 1: number of threads in the team: 2
 3  Level 2: number of threads in the team: 1
 4  Level 3: number of threads in the team: 1
 5  Level 2: number of threads in the team: 1
 6  Level 3: number of threads in the team: 1
 7
 8  % export OMP_NESTED=true
 9  % export OMP_MAX_ACTIVE_LEVELS=3
10  % ./nested-omp
11  Level 1: number of threads in the team: 2
12  Level 2: number of threads in the team: 2
13  Level 2: number of threads in the team: 2
14  Level 3: number of threads in the team: 2
15  Level 3: number of threads in the team: 2
16  Level 3: number of threads in the team: 2
17  Level 3: number of threads in the team: 2
```

Figure 12.10: **Results from our nested parallelism program** – We show two different runs with the program from Figure 12.9. In the first case with default settings for nested parallelism, the system does not create nested parallel regions. For the second run, we set OMP_NESTED to true and OMP_MAX_ACTIVE_LEVELS to 3 and we observed the expected execution of nested parallel regions.

Likewise, we can set the OMP_PROC_BIND and OMP_PLACES environment variables to a comma-separated list and assign items to levels using the same procedure.

Figure 12.11 illustrates thread affinity with nested parallelism in OpenMP. Assume our NUMA system has 2 sockets (i.e., two CPUs), 4 cores per socket, and 4 hardware threads per core. We will also assume we are running an application with two levels of nesting and that the parallel regions take their thread affinity and numbers of threads from environment variables (i.e., the code does not override the relevant internal control variables with clauses). Consider the following Environment Variable settings:

```
export OMP_PLACES=sockets,threads
export OMP_NUM_THREADS=2,4
export OMP_PROC_BIND=spread,close
```

When the program begins, there is one initial thread running on Core 0 on the first hardware thread. We encounter a parallel region and use the first values from OMP_NUM_THREADS and OMP_PROC_BIND (2 and spread). We create two threads, one

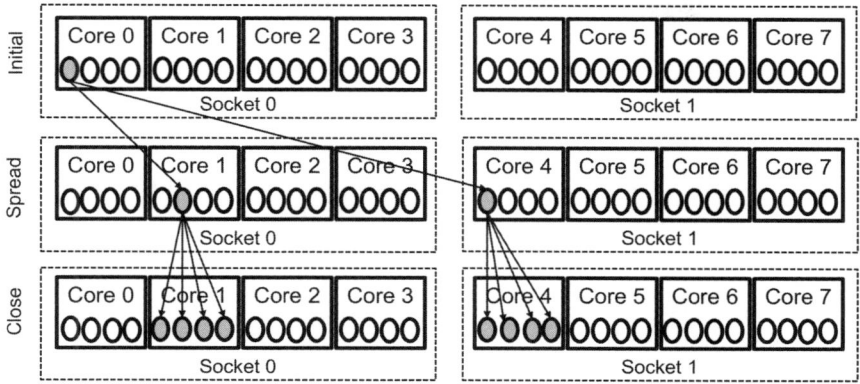

Figure 12.11: **Affinity with nested parallel regions** – Illustration of thread binding to logical CPUs on a system with 2 sockets, 4 cores per socket, and 4 hardware threads per core. OMP_PLACES=sockets,threads, OMP_NUM_THREADS=2,4, and OMP_PROC_BIND=spread,close. Note that in the middle layer (spread) the threads can migrate anywhere within their socket.

on each socket. Notice that the thread can run on any core and any hardware thread in the defined place, which being socket in this case, means they can run on any core in their respective sockets. After the first parallel region is created, the internal control variables go to the next values in the list: 4 for the number of threads and close for the processor binding. When the nested parallel region is encountered by each thread, they create 4 threads on the same core.

As with processor and data affinity, it is difficult to come up with simple rules for the best way to exploit nested parallel regions in OpenMP. Usually, it is best to use OMP_PROC_BIND=spread,close, to spread out at the higher levels of the hierarchy but to keep innermost parallel regions close together to better exploit cache locality.

### 12.1.3  Checking the Thread Affinity

When working with a NUMA system, you often need to verify that threads are being distributed as you expect. There are a number of system dependent ways to access this information. For example, with the Intel compilers you can set the environment variable KMP_AFFINITY=verbose to print detailed information about thread affinity. With the Cray compiler you can achieve a similar result with

CRAY_OMP_CHECK_AFFINITY=TRUE. Such approaches, however, are non-portable and at odds with OpenMP's focus on cross-platform portability.

In response to this problem, we added features to OpenMP 5.0 to expose thread affinity. This is done with a pair of environment variables:

- **OMP_DISPLAY_AFFINITY**: Set to TRUE or FALSE. Setting this to true will cause the system to display affinity information for all OpenMP threads when entering the first parallel region and when any thread affinity information changes in subsequent parallel regions.

- **OMP_AFFINITY_FORMAT**: Set to a string that defines the output affinity values that will be output and the format used when displaying them.

The string used with OMP_AFFINITY_FORMAT includes any substrings you choose plus output fields that define which thread information to output. A field has the form:

%[[[0].]size]type

where size defines the number of characters used for an output field and type indicates the information to output. The period indicates that the values are to be right-justified (the default is left-justified) and the 0 indicates that you want leading zeros to be included. A partial list of the available types (we leave off the ones that cover concepts we have not discussed) are defined in Table 12.1. Note that each of the types can be defined in two different ways: with single-letter/short-name or a longer and more descriptive long name. When you use the long name, it appears inside curly braces.

For a concrete example of how to use OMP_DISPLAY_AFFINITY, we will run the STREAM benchmark on the server node with logical CPU numbering from Figure 12.4. In Figure 12.12 we show all the details for how we built and ran this benchmark. The environment variables are particularly important:

- **OMP_DISPLAY_AFFINITY**: Set to true so the system will print information about the threads.

- **OMP_AFFINITY_FORMAT**: The format string using the short names for the types of information we want displayed.

- **OMP_PLACES**: The unit of binding for places where OpenMP threads will run. In this case, it is set to indicate binding should occur at the level of hardware threads.

Table 12.1: **The types used in OMP_DISPLAY_FORMAT**

| Long Name | Short Name | Meaning |
|---|---|---|
| nesting_level | L | The level in a hierarchy of nested parallel regions |
| thread_num | n | The thread's ID in its team |
| num_threads | N | The number of threads in a team |
| ancestor_tnum | a | The ID of the thread's parent |
| host | H | The name of the host machine the OpenMP program is running on |
| thread_affinity | A | The list of integers signifying the logical CPUs the threads can run on |

- **OMP_NUM_THREADS**: Default number of threads to use on parallel regions.

- **OMP_PROC_BIND**: Set to `spread` for the first run and `close` for the second.

We ran the benchmark (`./stream`) and filtered the output by thread number (the third field which we select by piping through `sort -k3`). Look closely at Figure 12.4 to follow how the threads were scheduled onto the node. For the first case (`spread`) as you add threads they go from the initial core to another core in the same socket, then to a core on the other socket, then to a core on that socket and then back to the first socket. This continues until all the threads are assigned. They are spread out among cores and across sockets. For the second case (`close`) the threads follow a different pattern. We start by adding a thread to the hardware thread on the starting core, then we go to another core in the same socket to assign the next thread after which we assign a thread to the next hardware thread in that same socket. This continues, filling hardware threads on each core before going to the next, and keeping all the threads in this case on one socket.

### 12.1.4   Summary: Thread Affinity and Data Locality

We live in a NUMA world. As the complexity of nodes in our servers increase, the need to directly address the NUMA features of a system becomes more and more important. This means understanding thread affinity relative to the specific features of your system and then using data in a way that maximizes locality.

```
1   $ icc −qopenmp −DNTIMES=20 −DSTREAM_ARRAY_SIZE=64000000 −c stream.c
2   $ icc −qopenmp −o stream stream.o
3   $ export OMP_DISPLAY_AFFINITY=true
4   $ export OMP_AFFINITY_FORMAT="Thrd Lev=%3L, thrd_num=%5n, thrd_aff=%15A"
5   $ export OMP_PLACES=threads
6   $ export OMP_NUM_THREADS=8
7   $ export OMP_PROC_BIND=spread
8
9   $ ./stream | sort −k3
10  <stream results omitted ...>
11  Thrd Lev=1  , thrd_num=0    , thrd_aff=0
12  Thrd Lev=1  , thrd_num=1    , thrd_aff=8
13  Thrd Lev=1  , thrd_num=2    , thrd_aff=16
14  Thrd Lev=1  , thrd_num=3    , thrd_aff=24
15  Thrd Lev=1  , thrd_num=4    , thrd_aff=1
16  Thrd Lev=1  , thrd_num=5    , thrd_aff=9
17  Thrd Lev=1  , thrd_num=6    , thrd_aff=17
18  Thrd Lev=1  , thrd_num=7    , thrd_aff=25
19
20  $ export OMP_PROC_BIND=close
21  $ ./stream |sort −k3
22  <stream results omitted ...>
23  Thrd Lev=1  , thrd_num=0    , thread_aff=0
24  Thrd Lev=1  , thrd_num=1    , thread_aff=32
25  Thrd Lev=1  , thrd_num=2    , thread_aff=2
26  Thrd Lev=1  , thrd_num=3    , thread_aff=34
27  Thrd Lev=1  , thrd_num=4    , thread_aff=4
28  Thrd Lev=1  , thrd_num=5    , thread_aff=36
29  Thrd Lev=1  , thrd_num=6    , thread_aff=6
30  Thrd Lev=1  , thrd_num=7    , thread_af=38
```

Figure 12.12: **Affinity format example** – We set the thread affinity format string and then ran the STREAM benchmark on the server-node with logical CPU numbering from Figure 12.4. We show two different executions of the STREAM benchmark: one with `OMP_PROC_BIND` set to `spread` and the other with `OMP_PROC_BIND` set to `close`.

Optimizing software for the NUMA features of a system is not straightforward. It also, almost by design, delves into the specific details of a particular system. You need to invest the time to understand your node architecture with tools such as "numactl -H". Explore your system with simple examples and the tools at your disposal to understand the system settings you expect to use for your actual applications.

A general recommendation is to have at least one process (e.g., a single MPI rank or OS process) per NUMA domain. Let OpenMP threads cover parallelism inside a NUMA domain keeping the data needed within the same NUMA domain. This reduces the impact of any errors in getting the first touch initializations correct across NUMA domain boundaries. Another general recommendation is to place threads far apart (`spread`) to take advantage of aggregated memory bandwidth then fork a nested parallel region `close` together for innermost work to maximize cache locality.

There is much more we could discuss about exploiting the NUMA features of a system. There are additional OpenMP runtime environment variables for managing nested parallel regions including:

- `OMP_THREAD_LIMIT`

- `OMP_NESTED`

- `OMP_MAX_ACTIVE_LEVELS`

There are also runtime library functions for thread affinity support, such as `omp_get_num_places`, `omp_get_place_proc_ids`, etc. OpenMP 5.0 added affinity support for explicit tasks. To learn more about OpenMP affinity, consult the book *Using OpenMP – The Next Step* [13].

## 12.2   SIMD

Modern CPUs include vector units. A single stream of vector instructions operates on specialized vector registers that hold multiple values. We call this the Single Instruction Multiple Data or SIMD execution model. Vector instructions fetch and decode a single instruction that is then applied to many data elements. They are fundamentally more energy efficient than scalar operations which decode an instruction for each set of operands. As the focus of computing has shifted from "raw performance" to "performance per watt", the energy efficiency of vector instructions has made them a key design element of modern microprocessors.

Vector units are characterized by the width of the vector registers. The original CPU vector units (the MMX instructions released in 1994) were only 64 bits wide. The width has grown over the years with 128 bits (Streaming SIMD Extensions or "SSE" in 1999), 256 bits (AVX in 2011), 512 bits (AVX-512 in 2013). As long as software can adapt and use these wide vector units, the width could climb even higher.

To understand how to exploit these vector units, we will transform a serial loop into one that uses vector instructions. We start the program we used in Chapter 4 to explore basic parallelism with threads: the numerical integration Pi program (Figure 4.5), and show a cleaner version again in Figure 12.13. We will work with an older and more straightforward vector instruction set: the SSE 4.2 instructions. The SSE 4.2 architecture extends the basic x86 architecture with 128-bit wide vector registers. Multiple variables with widths that add up to 128 bits can be accommodated by these instructions. The version of the program in Figure 4.5 works with float numbers which are 32 bits wide. We can pack 4 32-bit numbers into an SSE 4.2 vector register.

```
1   static long num_steps = 100000000;
2   float step;
3   int main()
4   {
5       int i;
6       float x, pi, sum = 0.0;
7
8       step = 1.0f / (double) num_steps;
9
10      for (i = 0; i < num_steps; i++) {
11          x = (i + 0.5f) * step;
12          sum += 4.0f / (1.0f + x * x);
13      }
14
15      pi = step * sum;
16  }
```

Figure 12.13: **Serial Pi program** –This program approximates a definite integral using the midpoint rule The loop iterations are independent other than the summation into sum. Note that we must explicitly represent all constants as floats to prevent internal operations from using double precision.

The basic idea of a vectorized program is to find loops that can be converted so the loop body executes with the vector instructions. In our case, we pack four floats into a vector register so we need each iteration of the vectorized loop to carry out four iterations from the scalar loop. This is called *unrolling the loop*. We show the four-fold, unrolled loop in Figure 12.14. To keep things simple, we assume that the number of iterations of the loop (num_steps) is evenly divided by 4. That way, we do not have any leftover iterations to deal with after we complete the unrolled loop.

The loop increment is changed so each pass through the loop handles 4 iterations:

```
for (i = 1; i <= num_steps; i = i + 4)
```

The loop body is modified to handle four iterations. Instead of setting a single value for x, we set four values incremented to cover the values from four iterations, x0, x1, x2, and x3. We then compute the integrand for each of the x values and sum them together:

```
sum = sum + 4.0f * ( 1.0f/(1.0f+x0*x0) + 1.0f/(1.0f+x1*x1)
                   + 1.0f/(1.0f+x2*x2) + 1.0f/(1.0f+x3*x3) );
```

An unrolled loop reduces overhead. In our example, we go through the loop only one quarter as many times. We are not overly concerned with performance at this point, however. Our goal for unrolling the loop is to prepare our program to explicitly add vector instructions.

```
1    static long num_steps = 100000000;
2    float step;
3    int main ()
4    {
5        int i;
6        float xo, x1, x2, x3, pi, sum = 0.0;
7
8        step = 1.0f / (double) num_steps;
9
10       for (i = 1; i <= num_steps; i = i+4) {
11           x0 = (i - 0.5f) * step;
12           x1 = (i + 0.5f) * step;
13           x2 = (i + 1.5f) * step;
14           x3 = (i + 2.5f) * step;
15           sum = sum + 4.0f * (1.0f/(1.0f+x0*x0) + 1.0f/(1.0f+x1*x1) \
16                       + 1.0f/(1.0f+x2*x2) + 1.0f/(1.0f+x3*x3));
17       }
18       pi = step * sum;
19       printf("pi = \%lf, \%ld steps\n ",pi, num_steps);
20   }
```

Figure 12.14: **Serial Pi program with loops unrolled by 4** – Numerical integration to estimate Pi. We assume the number of steps is evenly divided by 4 just to keep the program simpler.

To transform our loop-unrolled version of the program into a vectorized program, we need to add the explicit vector instructions. We can do this without switching to

assembly code by using a common set of vector intrinsics. The style of programming
with intrinsics is similar to assembly coding in that you fill registers and apply
operations on those registers to produce values in other registers. With vector
intrinsics, however, we do this with a fixed API of C functions. In particular, we
will use the common vector intrinsics for SSE defined in x86intrin.h [4].

A detailed description of the instructions we will be using from x86intrin.h
would go well beyond the scope of this discussion. Instead, we'll give you a high level
understanding: enough to follow the basic idea of our vectorized code in Figure 12.15.
The SSE vector registers are 128 bits wide. To indicate the case where we pack 4
32-bit, float values into each register, the registers are of type __mm128. We declare
the registers we will use in the program in lines 12 through 19. We assign a set
of explicit literals (in this case a "ramp" from 0.5 to 3.5) to a register with the
following statement:

```
__m128 ramp = _mm_setr_ps(0.5, 1.5, 2.5, 3.5);
```

and load each of the four floats with the value of an input scalar variable:

```
__m128 vstep = _mm_load1_ps(&step);
```

We then replace the body of our unrolled loop with vector instructions that operate
on the vector registers we have established. The algorithm is straightforward
though if you are not used to programming in assembly code, it may look confusing.
Basically, we chain instructions together so they flow from our initial vector register,
through intermediate values and ends with assignment to our destination register.
Consider the following statement from line 24:

```
xvec = _mm_mul_ps(_mm_add_ps(eye,ramp), vstep);
```

We add our previously defined "ramp" to the loop control index (_mm_add_ps) and
then stream that result into a multiply operation (_mm_mul_ps) to produce our
vector of x values. If you look at our unrolled loop in Figure 12.14 you can hopefully
see that at the conclusion of that vector operation we have packed the values x0,
x1, x2, and x3 into the single vector register xvec. We continue in this same vein
to produce a vector register (denom) containing four values of the denominator in
our integrand $(1.0 + x^2)$:

```
denom = _mm_add_ps(_mm_mul_ps(xvec, xvec), one);
```

We close the body of the loop by dividing 4.0 by the denominator (using `_mm_div_ps`) and then summing the result into the **sum** register:

```
sum = _mm_add_ps(_mm_div_ps(four, denom), sum);
```

When the loop has finished we map the values from the vector registers back into a regular array and them sum those elements together to get the final result:

```
_mm_store_ps(&vsum[0], sum);
pi = step * (vsum[0] + vsum[1] + vsum[2] + vsum[3]);
```

Since our ultimate goal is to combine multithreaded programming and vectorized, SIMD code we have one more version of the vectorized program to consider. In Figure 12.16 we show how to mix OpenMP and explicitly vectorized code. We use a mixture of techniques we learned with the SPMD pattern and loop-level parallelism. We create our parallel region and then declare the set of vector registers. This time, however, we are creating a set of vector registers for each thread. We then use an OpenMP worksharing-loop to give different blocks of SIMD instructions to each thread. On lines 40 and 41 when we extract the floats packed into the **sum** register and combine them to create the final **sum**, we do this on a per-thread basis and place the result in an array indexed by the threadID (just as we did in Figure 4.6 when working with our first SPMD Pi program).

The vast majority of programmer will never write explicitly vectorized code. To most programmers, vectorization is done by a compiler. For most compilers, you set the optimization flag to level 3 (-O3) to tell the compiler to aggressively optimize the code which includes automatically vectorizing the program. It is important, however, to understand at a high level what a compiler does to your code when it generates vector code. This is why we took the time to explain loop unrolling and the use of vector intrinsics.

Compiler technology is truly amazing. From reordering instructions to maximize throughput to automatically finding ways to exploit vector units, a modern compiler saves us from the hard work of optimizing code by hand. As impressive as the technology is, however, a compiler first and foremost is required to produce correct answers. Hence, when in doubt about the semantic impact of any code restructuring, the compiler will leave the code alone. The result is that compilers fail to

```
1
2   #include <x86intrin.h>
3   static long num_steps = 100000000;
4   float scalar_four = 4.0f, scalar_zero = 0.0f; scalar_one = 1.0f;
5   float step;
6   int main ()
7   {
8       int i;
9       float xo, x1, x2, x3, pi, sum = 0.0;
10      step = 1.0f/(double) num_steps;
11
12      __m128 ramp  = _mm_setr_ps(0.5, 1.5, 2.5, 3.5);
13      __m128 one   = _mm_load1_ps(&scalar_one);
14      __m128 four  = _mm_load1_ps(&scalar_four);
15      __m128 vstep = _mm_load1_ps(&step);
16      __m128 sum   = _mm_load1_ps(&scalar_zero);
17      __m128 xvec;
18      __m128 denom;
19      __m128 eye;
20
21      for (i = 0; i < num_steps; i = i + 4){
22          ival  = (float) i;
23          eye   = _mm_load1_ps(&ival);
24          xvec  = _mm_mul_ps(_mm_add_ps(eye,ramp), vstep);
25          denom = _mm_add_ps(_mm_mul_ps(xvec,xvec), one);
26          sum   = _mm_add_ps(_mm_div_ps(four,denom), sum);
27      }
28      _mm_store_ps(&vsum[0], sum);
29
30      pi = step * (vsum[0] + vsum[1] + vsum[2] + vsum[3]);
31  }
```

Figure 12.15: **Pi program using SSE vector intrinsics** – Numerical integration to estimate Pi. We assume the number of steps is evenly divided by 4 just to keep the program simpler.

vectorize many loops. The utilization of vector instructions in modern programs is, unfortunately, quite low.

We need to help the compilers. When we know that loops are independent and do not have loop-carried dependencies, we can force the compiler to vectorize the loop. We do this with the simd construct on a loop. For example, in Figure 12.17 just before the loop over steps we added the directive:

```
#pragma omp simd private(x) reduction(+:sum)
```

```
1   #include <omp.h>
2   #include <x86intrin.h>
3   static long num_steps = 100000000;
4   #define MAX_THREADS 4
5   double step;
6   int main ()
7   {
8       int i;
9       float local_sum [MAX_THREADS];
10      float xo, x1, x2, x3, pi, sum = 0.0;
11      step = 1.0f / (double) num_steps;
12
13      for (k = 0; k < MAX_THREADS; k++) local_sum[k] = 0.0;
14
15      #pragma omp parallel num_threads(4)
16      {
17          int i, ID = omp_get_thread_num();
18          float scalar_one = 1.0, scalar_zero = 0.0;
19          float ival, scalar_four = 4.0;
20          float vsum[4];
21
22          __m128 ramp  = _mm_setr_ps(0.5, 1.5, 2.5, 3.5);
23          __m128 one   = _mm_load1_ps(&scalar_one);
24          __m128 four  = _mm_load1_ps(&scalar_four);
25          __m128 vstep = _mm_load1_ps(&step);
26          __m128 sum   = _mm_load1_ps(&scalar_zero);
27          __m128 xvec;
28          __m128 denom;
29          __m128 eye;
30
31          // unroll loop 4 times ... assume num_steps\%4 = 0
32          #pragma omp for schedule(static)
33              for (i = 0; i < num_steps; i = i + 4) {
34                  ival  = (float)i;
35                  eye   = _mm_load1_ps(&ival);
36                  xvec  = _mm_mul_ps(_mm_add_ps(eye,ramp), vstep);
37                  denom = _mm_add_ps(_mm_mul_ps(xvec,xvec), one);
38                  sum   = _mm_add_ps(_mm_div_ps(four,denom), sum);
39              }
40          _mm_store_ps(&vsum[0],sum);
41          local_sum[ID] = step * (vsum[0]+vsum[1]+vsum[2]+vsum[3]);
42      }
43
44      for (k = 0; k < MAX_THREADS; k++) pi += local_sum[k];
45
46      pi = step * (vsum[0]+vsum[1]+vsum[2]+vsum[3]);
47  }
```

Figure 12.16: **A multithreaded and vectorized Pi program** – This program carries out a numerical integration to estimate Pi. We assume the number of steps is evenly divisible by 4 and that we got 4 threads just to keep the program simple.

This directive asserts to the compiler that it is safe to transform the loop into vectorized code. Unlike the automatic vectorization we use routinely, the compiler will not analyze the loop and only carry out the vectorization if it can prove it is safe to do so. The simd construct is explicit and forces the compiler to generate vector instructions for the loop. The simd construct can be combined with the worksharing-loop construct to create a composite construct. We show this composite construct combined with a parallel construct in Figure 12.18. The OpenMP compiler will decompose the loop into contiguous chunks scheduled to execute on each thread and then, within each chunk, it will unroll the loop and transform the body of those unrolled loops so they use SIMD instructions. The decomposition of the loop iterations into chunks for multithreaded execution takes precedence over the vectorization enabled by the simd clause. If the chunk size from the parallel-loop decomposition is not evenly divisible by the width of the blocks needed by the simd clause, then the vectorization could be compromised.

```
1   #include <omp.h>
2   static long num_steps = 100000000;
3   double step;
4   int main ()
5   {
6       int i;
7       float x, pi, sum = 0.0;
8
9       step = 1.0f / (double) num_steps;
10
11      #pragma omp simd private(x) reduction(+:sum)
12          for (i = 0; i < num_steps; i++) {
13              x = (i + 0.5f) * step;
14              sum += 4.0f / (1.0f + x * x);
15          }
16
17      pi = step * sum;
18  }
```

Figure 12.17: **OpenMP program to vectorize the Pi program** – The simd clause directs the compiler to explicitly vectorize the program. As with many OpenMP features, this clause asserts to the compiler that it is safe to vectorize the code and it will do so, even if there are loop-carried dependencies that should prevent vectorization.

Up to this point, we have not discussed the performance of these versions of the Pi program. We present these in Table 12.2. These are averaged times over

```
 1  #include <omp.h>
 2  static long num_steps = 100000000;
 3  double step;
 4  int main ()
 5  {
 6      int i;
 7      float x, pi, sum = 0.0;
 8
 9      step = 1.0f / (double) num_steps;
10
11      #pragma omp parallel for simd private(x) reduction(+:sum)
12      for (i = 0; i < num_steps; i++) {
13          x = (i + 0.5f) * step;
14          sum += 4.0f / (1.0f + x * x);
15      }
16
17      pi = step * sum;
18  }
```

Figure 12.18: **OpenMP program to multithread and vectorize the Pi program** – This is a familiar "parallel for" approach to solving the problem but we have added one additional clause: a **simd** clause for explicit vectorization.

50 executions of the Pi loop. The averages were stable over multiple runs of the program. The unrolled loop has reduced loop overhead and runs slightly faster. Notice that since the optimization level used in compiling these program includes vectorization, both the *Base-float* and *Unroll-4* results include autovectorization. We confirmed this by analyzing the optimization report generated by the compiler.

It is interesting to compare those autovectorized results to the explicit vectorization in the row labeled as *SSE*. This is a common result: autovectorization usually fails to match the performance from a hand vectorized program. Compilers are required to return correct results and therefore they are very conservative in the transformations they are willing to apply to a program.

The addition of threads in the row labeled *SSE+OMP-par* produces a substantial speedup. Based on experience with this Pi program from earlier in the book, if we had more steps in the integration so the total run time was larger relative to the parallel loop overhead, we would expect the speedup to be considerably better.

Finally we come to the new **simd** construct introduced in this chapter. The row labeled *OMP-SIMD* shows the vectorization resulting from the **simd** construct. Notice that it is on par with the result from the autovectorized result and does not

Table 12.2: **Run times in seconds for our numerical integration programs**
– We ran these programs with 8388608 steps 50 times and averaged the run times. We used
4 threads on a dual-core system with 2 threads per physical core (i.e., hyperthreading was
enabled). Code was compiled with the Intel® icc compiler version 18.0.3 with optimization
at −O3. The system was an Apple® Macbook Air®, 2.2 GHz Intel® Core™ i7 with 8
MB of DDR3 RAM at 1600 MHz.

| Case (Figure#) | Pi | Ave time (seconds) |
|---|---|---|
| Base-float (12.13) | 3.140426 | 0.004849 |
| Unroll-4 (12.14) | 3.141240 | 0.004266 |
| SSE (12.15) | 3.139504 | 0.003380 |
| SSE+OMP-par (12.16) | 3.140708 | 0.002278 |
| OMP-SIMD (12.17) | 3.140426 | 0.004930 |
| OMP-par-SIMD (12.18) | 3.141475 | 0.002651 |

perform as well as the explicit SSE vectorization. The loop used in this chapter
is quite simple and it is not surprising that the explicit `simd` construct was not
able to show a greater benefit. The final result is when we use the composite
`for/simd` construct combined with a `parallel` construct. This result shows a nice
overall improvement in performance, though of course it does not match the parallel
construct combined with SSE.

There is much more we could discuss about the `simd` construct in OpenMP.
The `simd` construct uses the familiar data environment clauses (`private` and
`lastprivate`), the `collapse` clause, and the `reduction` clause. Additional clauses
on the construct control assumptions that can be made about patterns of dependencies in the loop (`safelen`, `simdlen`, and `linear`). Other constructs provide
information to the compiler about assumptions that can be made about how variables
inside the simd loop access memory (`aligned` and `nontemporal`). To learn more
about the `simd` construct, consult the book *Using OpenMP – The Next Step* [13].

## 12.3   Device Constructs

Starting with version 4.0, OpenMP adopted a *host-device model*. When you launch
an OpenMP program, it starts running on the *host*. Much of what we have covered
in this book focus on how to use multiple threads to optimize performance on the
host. Attached to the host are one or more *devices*. These can be other CPUs but
typically they are GPUs.

When GPUs became programmable devices suitable for more than rendering images, they introduced a new execution model. There is no perfect name for this model, but it is usually called "Single Instruction Multiple Thread" or SIMT. The name is confusing in that the "threads" in SIMT are not the same as "threads" in OpenMP. The name "SIMT" has stuck, however, so we will follow common practice and use it.

We will describe the core ideas behind the SIMT execution model with the code in Table 12.3. Consider a traditional loop oriented view of a simple computation as presented on the left side of the table. This is a classic data parallel operation. The computation can be described as "for each index i ranging from 0 to (n-1), add the product of a[i] and b[i] to c[i]". The SIMT model transforms this data parallel operation into execution on an attached device optimized for data parallel execution. The model assumes two programs: one running on a host (a CPU) and the other runs on a device (typically a GPU). The process of running a program that follows the SIMT model breaks down into the following steps:

- The loop over i is replaced with an index space called a "grid" (in CUDA) or an NDRange (in OpenCL)[2].

- The body of the loop is turned into a special type of function called a *kernel*. We show the kernel corresponding to our example data parallel loop in the right side of Table 12.3. Since we emphasize vendor-neutral programming, we use OpenCL. Variables marked with the qualifier *global* come from the address space associated with the attached device. The function `get_global_id(0)` returns the index for a point in the NDRange on which the kernel will execute.

- Data objects (the arrays a, b, and c in our example) are copied from the host onto the device.

- A kernel and an associated NDRange is enqueued for execution on a host and then "offloaded" to run on an attached device.

- An instance of the kernel runs at each point in the index space. This instance is called a "work-item" in OpenCL or a "thread" in CUDA.

- The computation completes and any results are copied back to the host.

---

[2]OpenCL [11] is an industry standard language for programming data parallel devices such as GPUs. Think of it as a vendor-neutral instantiation of the basic programming model defined in CUDA® from Nvidia®.

The SIMT concept is straightforward. You line up data and kernels around the index space (i.e., the NDRange), organize work-items into blocks that execute independently (to hide memory access latencies), and stream computation through the device to hopefully generate results at high throughput. The problem is that with traditional GPU languages such as OpenCL or CUDA, you need to write two programs: one for the host and one for the device. In the case of a vendor-neutral programming language such as OpenCL, few assumptions can be made about the device so the host program can be dozens of lines of code even for the simple case we show in Table 12.3.

Table 12.3: **The SIMT model** – Replace a loop with a kernel function. An instance of the kernel runs at each point in an index space defined by the loop, in this case a one dimension index space from 0 to n−1.

| Traditional Loop code | OpenCL code for a GPU |
|---|---|
| void mul (const int n,<br>  const float *a,<br>  const float *b,<br>  float *c)<br>{<br> int i;<br> for (i=0; i<n; i++)<br>  c[i] += a[i]*b[i];<br>} | kernel void mul (<br>  global const float *a,<br>  global const float *b,<br>  global float *c)<br>{<br> int id = get_global_id (0);<br><br> c[id] += a[id]*b[id];<br>} |

Transforming a loop-oriented view of a data parallel computation into one based on the SIMT model is straightforward to express as a sequence of basic rules. The application of these rules can be automated. Hence, in OpenMP version 4.0, we added directives to support offloading computations onto a device. We show a program for our data parallel vector multiplication example in Figure 12.19.

The program starts on the host as a regular CPU program. The `target` directive and its associated structured block (the single for loop in this case) define the *target region* offloaded for execution on the device. The `target` directive also causes data to move onto the device, which in the case of Figure 12.19 would be the arrays a, b, and c as well as the scalars i and N. This data will be copied from the device back onto the host when the target region completes execution.

OpenMP is designed for devices ranging from other CPUs, GPUs, and even FPGAs. Hence, the directive paired with the `target` directive can be quite complicated.

```
1   #include<omp.h>
2   #include<stdio.h>
3   #define N 1024
4   int main()
5   {
6       float a[N], b[N], c[N];
7       int i;
8
9   // initialize a, b, and c (code not shown)
10
11  #pragma omp target
12  #pragma omp teams distribute parallel for simd
13      for (i = 0; i < N; i++)
14          c[i] += a[i] * b[i];
15  }
```

Figure 12.19: **OpenMP program for elementwise multiplication of vectors
on a GPU** – Default data movement moves the vectors a, b, and c onto the device before
the computations starts and back onto the host (the CPU) when the computation has
completed.

For a GPU, the best approach is to follow the **target** directive with the following
directive and its associated loop:

```
#pragma omp teams distribute parallel for simd
```

This directive will turn the body of the subsequent loop into a kernel and construct
an index space (i.e., an NDRange) from the loop control indices. It will create a
team of threads for each computational unit on a device (such as the streaming
SIMD units on a GPU) and distribute blocks of work-items for the loop iterations
among the teams of threads. In other words, it will do essentially the identical
SIMT execution strategy we described earlier for OpenCL or CUDA.

Data movement has a major impact on performance. Typically, the data moves
from the host to the device over a PCI link which is much slower than the speed of
moving data directly from memory. Hence, we have added to the **target** directive
a clause to define explicit data movement. We show this **map** clause and how it
is used in Figure 12.20. In this case, the map clause stipulates that the arrays a
and b are to be moved[3] **to** the device at the beginning of the computation. The

---

[3]Words such as "moved" and "copied" imply that data is physically copied between the host and
a device. In terms of the logical meaning of the map clauses, that view is consistent. OpenMP

array c is to be moved onto the device before the computation begins but when the computation has finished, the contents of the array are to be moved back onto the host (i.e., **from** the device to the host). As an additional feature of Figure 12.20, we also show how to use array sections to define the data to be moved which is needed when the arrays are defined by pointers.

```
1   #include<omp.h>
2   #include<stdio.h>
3   #define N 1024
4   int main()
5   {
6       float *a, *b, *c;
7       int i;
8
9       a = (float*) malloc(N * sizeof(float);
10      b = (float*) malloc(N * sizeof(float);
11      c = (float*) malloc(N * sizeof(float);
12
13  // initialize a, b, and c (code not shown)
14
15  #pragma omp target map(to:a[0:N],b[0:N]) map(tofrom:c[0:N])
16  #pragma omp teams distribute parallel for simd
17      for (i = 0; i < N; i++)
18          c[i] += a[i] * b[i];
19  }
```

Figure 12.20: **Explicit data movement with the target directive** – The map clause controls movement of data from the host **to** a device or **from** the device onto the host. When working with pointers to arrays, you need to use array sections to define precisely which data to move.

In Figure 12.21 we add one more complication to our problem. We now have two target regions that compute an array c and then use that to compute d. Map clauses on each target region manage the data movement. However, we move the array a onto the device twice, once for each kernel. The array c is moved onto the host at the end of the first target region and then right back onto the device for the second target region.

---

is very careful, however, to *not* require actual physical movement of the data. When a device shares an address space with the host, an implementation may "transfer ownership" without the overhead of actually moving the data.

```
 1   #include <omp.h>
 2   #include <stdio.h>
 3   #define N 1024
 4   int main()
 5   {
 6       float *a, *b, *c, *d;
 7       int i;
 8
 9       a = (float*) malloc(N * sizeof(float);
10       b = (float*) malloc(N * sizeof(float);
11       c = (float*) malloc(N * sizeof(float);
12       d = (float*) malloc(N * sizeof(float);
13
14   // initialize a, b, c, and d (code not shown)
15
16   #pragma omp target map(to:a[0:N],b[0:N]) map(tofrom:c[0:N])
17   #pragma omp teams distribute parallel for simd
18       for (i = 0; i < N;i++)
19           c[i] += a[i] * b[i];
20
21   #pragma omp target map(to:a[0:N],c[0:N]) map(tofrom:d[0:N])
22   #pragma omp teams distribute parallel for simd
23       for (i = 0; i < N; i++)
24           d[i] += a[i] + c[i];
25   }
```

Figure 12.21: **Multiple target regions** – The map clause controls movement of data from the host **to** a device or **from** the device onto the host. When working with pointers to arrays, you need to use array sections to define precisely which data to move.

We need to manage data at the level of the device, not at the level of individual target regions. We do this with a construct called a "target data region". This lets us define a data region once that is used across multiple target regions. We show an example of this construct in Figure 12.22. The arrays a, b, c, and d are copied from the host onto the device only once before any of the target regions execute. Then at the end of the target data region, the array d is copied back to the host.

There is much more we could cover about the device constructs in OpenMP. As you can imagine, given the diversity of possible devices and the wide range of algorithms possible for data parallel devices, this topic can be very complex. The basic concept of SIMT, however, and how it maps onto OpenMP at a high level is straightforward. By tying the index space at the heart of SIMT to loop nests and loop-bodies to kernels, the expression of parallelism is generic and maps directly

```
1    #include<omp.h>
2    #include<stdio.h>
3    #define N   1024
4    int main()
5    {
6        float *a, *b, *c, *d;
7        int i;
8
9        a = (float*)malloc(N*sizeof(float);
10       b = (float*)malloc(N*sizeof(float);
11       c = (float*)malloc(N*sizeof(float);
12       d = (float*)malloc(N*sizeof(float);
13
14   // initialize a, b, c, and d (code not shown)
15
16   #pragma omp target data map(to:a[0:N],b[0:N],c[0:N]) map(tofrom:d[0:N])
17   {
18       #pragma omp target
19       #pragma omp teams distribute parallel for simd
20       for (i = 0; i < N; i++)
21           c[i] += a[i] * b[i];
22
23       #pragma omp target
24       #pragma omp teams distribute parallel for simd
25       for (i = 0; i < N; i++)
26           d[i] += a[i] + c[i];
27   }
28
29   // continue in the program but only using d (not c)
30
31   }
```

Figure 12.22: **Target Data Region** – A single target data region manages data at the level of a device. It persists and is used between multiple target constructs.

onto a wide range of hardware without the need for machine specific constructs. For more details about working with OpenMP devices, consult the book *Using OpenMP – The Next Step* [13].

## 12.4   Closing Comments

OpenMP is over 20 years old. Over most of its "life", the language has focused on SMP systems. The OpenMP Common Core, with its focus on SMP systems, reflects that history. As we discuss at length in this chapter, however, actual hardware is anything *but* SMP. The SMP model works since those implementing OpenMP compilers and the supporting runtime systems have done an excellent job of making multithreaded programs run fast. At some point, however, you will encounter situations where you need to go beyond SMP.

For OpenMP, there are three hardware trends we have embraced: NUMA, vector units, and attached devices (such as GPUs). We covered each of these in this chapter. Most of our time was spent on NUMA systems since they are quite common. Even a basic multicore CPU can benefit from treating it as a NUMA system.

Vector units, addressed with the SIMD construct, are also quite common. Compilers attempt to automatically exploit vector instructions so programmers often do not even consider directly programming them. As the width of these vector units grows, the need to effectively exploit them will grow as well. We made a difficult choice in the vectorization topics we choose to cover. The SIMD construct and associated clauses are well covered in the book *Using OpenMP – the Next Step* [13]. What is not well-covered in that book is the actual transformations a compiler must do to vectorize code. Therefore, we choose to describe those transformations rather than all the various clauses you might use with the `simd` construct.

We closed with the device constructs. GPUs are growing in popularity. For data parallel algorithms and problems that fit in the GPU's memory, they can be great. The central idea is to treat the GPU as a throughput-optimized device and stream work through the GPU. The central concept for OpenMP and the GPU is to treat a loop-nest as defining an index space and then run a function (a "kernel") at each point in that index space. It is a simple, but powerful idea that maps beautifully onto OpenMP.

# 13 Your Continuing Education in OpenMP

Depending on how you count them, the latest OpenMP specification (version 5.0) has 43 directives, 45 clauses, 21 runtime environment variables, 68 runtime library routines, and over 30 combined/composite directives. A complete reference guide covering all of those language features would be over 1000 pages! It would be an absurd book to write.

OpenMP is a growing language. On average, a new Specification is released every two years. Given how long it takes to write a book, a complete reference guide covering the whole language would be out of date shortly after it was printed. OpenMP is moving too fast for any book to keep up.

Therefore, OpenMP books focus on core concepts and fundamental design patterns. If you need to understand the full breadth of OpenMP and the latest directives and clauses, you have to go back to the source of OpenMP. You need to work directly with the resources provided by the OpenMP Architecture Review Board.

In this chapter, we introduce those resources. Our goal is to help you understand the resources available to help you become a more effective OpenMP programmer. One of those resources is the OpenMP Specification itself. A Specification is written for people who write compilers that support OpenMP, not application programmers. Hence, the Specification is full of complex jargon and obscure details that challenge even the most experienced application programmers. In this chapter, we will explain the specialized jargon used in the Specification so you can read it and extract the information you need.

## 13.1 Programmer Resources from the ARB

The OpenMP Architecture Review Board (ARB) communicates with the programmer community primarily through the OpenMP web site:

> https://www.openmp.org

On this web site you'll find news and events, background information about the ARB, and of course a Blog. As programmers, however, there are three items on the web site that are essential.

- OpenMP Specifications

- OpenMP Examples

- OpenMP Reference Guides

The Specifications are found under the web site's "Specifications" tab. Included with each Specification is the OpenMP Examples document. The Examples take a year or longer to be completed after the release of a new Specification, but eventually there is one Examples document for each Specification.

The Examples document is not a formal part of the standard. To most programmers, however, it is more important than the Specifications. While we will describe the OpenMP Specifications later in this chapter and help you learn what you need to productively work with them, programmers do not typically learn by reading Specifications. Programmers learn by looking at examples.

The contents of the Examples document covers most of the constructs in a Specification. Emphasis is given to the newer and more confusing constructs. The set of examples also include cases that expose subtle *issues* in OpenMP. For example, on page 62 of the OpenMP 4.5 Examples document, a particularly challenging "deadlock" problem is presented. We reproduce this code in Figure 13.1.

```
1   void work()
2   {
3      #pragma omp task   //task 1
4      {
5
6          #pragma omp task      //task 2
7          {
8
9                   #pragma omp critical // Critical region 1
10                    {/* do work here */}
11          }
12          #pragma omp critical // Critical Region 2
13          {
14
15             #pragma omp task // task 3
16              {/* do work here */}
17          }
18      }
19   }
```

Figure 13.1: **A subtle deadlock with tasks**: – This is the tasking.9.c example from the OpenMP 4.5 Examples document. This function can deadlock if the thread suspends task 1 to begin work on task 2.

The OpenMP 4.5 Specification defines scheduling rules for tasks and states that at points where a thread is allowed to suspend and activate tasks (a so-called "task scheduling point"), a program must assure that mutual exclusion constructs such as locks or critical sections are not held across a task scheduling point. Look closely at the code in Figure 13.1. The thread executing Task 1 creates a new explicit task (Task 2). It then enters a critical region inside of which it creates an additional explicit task (Task 3). Creation of an explicit task defines a task scheduling point, that is, a point in program execution where a thread can change which tasks are scheduled for execution. It is possible that at the task-scheduling point for Task 3, the runtime system could suspend Task 1 and execute Task 2. The critical construct is held by Task 1 so the program will deadlock at the critical construct inside Task 2.

This is a subtle bug that could easily confuse experienced OpenMP programmers. It is with considerable foresight that the ARB thought to put these sort of "subtle error" examples in the Examples document.

The examples are a key learning resource, but when coding a new OpenMP program, the challenge is less about learning and more about memorization. How can you keep the detailed syntax of all the constructs straight or remember which clauses are allowed on which directives? Under the Resources tab is a menu. One of the items on that menu is called "Reference Guides". Reference Guides summarize OpenMP syntax for each of the items in an OpenMP Specifications. We highly recommend that you download the Reference Guide for the version of OpenMP you are using and keep it with you as you write code.

## 13.2   How to Read the OpenMP Specification

The OpenMP Common Core is a subset of the OpenMP 3.0 Specification released in May of 2008. As we were writing this book, OpenMP 5.0 was the latest Specification. It was released in November of 2018. A great deal has happened to OpenMP in the 10 years between version 3.0 and 5.0. We added capabilities for addressing the needs of nonuniform memories. We expanded the range of tasking algorithms that can be addressed. We added support for vectorization with the `simd` construct. And most dramatically, we moved beyond multithreading and added directives for programming GPUs with OpenMP.

Regardless of all these new capabilities, however, the specialized jargon of OpenMP has remained much the same. In order to incorporate tasks into OpenMP, we had to change the language we used to describe the OpenMP constructs. These changes put tasks at the core of everything that happens in OpenMP. This lets us define behaviors of the data environment and the execution model in a consistent way

in terms of tasks. To make this work and support strict consistency across the
language, we had to add implicit parallel regions and other concepts that seem
strange when first encountered. The most important job of this section is to explain
this terminology so it will not confuse you when you encounter it in the Specification.

We will cover the jargon of OpenMP by explaining the language from the beginning,
but unlike elsewhere in the book we will use all the detailed jargon you will find in
the Specification. As you read this description of OpenMP, remind yourself that
you have been exposed to all these concepts. If the words seem confusing, pause
and remember that nothing we are describing is new to you.

### 13.2.1  OpenMP with All the Formal Jargon

OpenMP is a programming language for writing parallel applications. The hardware
is composed of one or more devices. One device is the *host device*. A program
execution begins in the host device. OpenMP up until version 4.0 only had the
host device. To be interesting, this device had to be a multiprocessor system such
as multi-core CPU. From OpenMP 4.0 onward, the language supports additional
devices. An OpenMP program can offload work onto these additional devices using
target directives. We will not consider offloading onto target devices further in this
discussion. We only mention them so you can appreciate the concept of a device in
OpenMP.

The OpenMP programming language is an extension of a base language. The
extensions are expressed through a set of directives and runtime library routines
defined by the OpenMP Specification. The base languages currently supported for
OpenMP are C, C++, and Fortran. Directives specify the behavior of the OpenMP
program and are `pragmas` in C/C++ and a specific form of comment statements in
Fortran.

Directives can be *declarative* or *executable*. A declarative directive appears with
the declarative statements in a program. It influences features of data or how a
function is compiled. A good example is the `threadprivate` directive introduced
in Section 10.2.1. Other directives are executable. As the name implies, these
directives modify the behavior of a program's execution. The `parallel` directive is
an example of an executable directive. An executable directive that is not associated
with any code is called a `stand-alone` directive. The `barrier` directive is the classic
example of a stand-alone directive.

The executable directives are fundamental to any useful parallel execution in an OpenMP program. An executable directive plus the code associated with that directive is called a *construct*. This code is almost always a structured block which includes for-loops, single statements, or a block of code with one point of entry at the top and one point of exit at the bottom. The code in the construct resides in the same compilation unit as the directive. It is hence often said to be in the lexical extent of the executable directive.

The other key idea to understand about executable directives are *regions*. A region is all the code executed by a construct. It includes code in the lexical extent of the directive but also code within functions executed by the construct. For example, for a team of threads created by a parallel construct, each thread executes the code within the *parallel region.*

We have now covered the hardware (the host device and any other attached devices) and how OpenMP interacts with the source code of a program (written with a base language and directives) leading to constructs and regions. Other than the SIMD and target constructs, OpenMP is concerned with threads that execute on a multiprocessor system. A thread is an execution entity. It has a program counter, its own private memory implemented as a stack, and some static memory attached to the thread (threadprivate memory). Notice, however, that in talking about a thread, we do not define the work it carries out. That work is captured by the concept of a task.

A task is a specific instance of executable code and its data environment. Tasks are executed by threads. We define all the work of a program in terms of tasks. By doing so, the tasks give us a common way to describe how code executes and interacts with a data environment. A task is an explicit task if it is created by a task construct. We call a task created by any other construct or one implied by program execution directly "an *implicit task*".

As an example of how to use this jargon in practice, we will discuss how a program executes when it is started on the host device. An implicit parallel region surrounds the whole program, where we use the term "implicit" since that parallel region is not created by a parallel construct. It is an inactive parallel region since it was not generated by a parallel construct and therefore is not executed in parallel by a team of threads. It only exists as a concept in OpenMP so we can define the elements of OpenMP consistently whether they are executed by a team of threads or not.

The implicit parallel region runs on the initial thread and executes the initial task. We see this in the first three points in Figure 13.2. At some point in the

OpenMP program the initial thread encounters a parallel construct. The initial task forks a team of threads and then suspends. In other words, the initial task becomes inactive and each team of threads runs an implicit task defined by the code within a parallel region. This is shown in points 4 through 7 in Figure 13.2. As implicit tasks complete their work, the threads associated with those tasks wait at the barrier. All tasks, both implicit and explicit (i.e., those created with a task construct), must terminate before the barrier is satisfied. At that point the threads in the team are terminated and the initial thread continues to run the initial task.

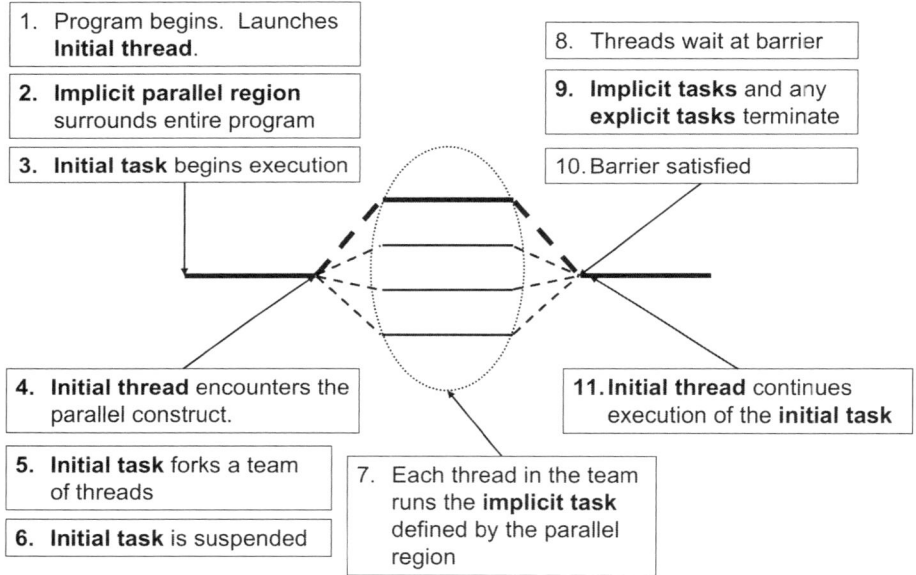

Figure 13.2: **Terminology associated with execution of a parallel region** – A program begins as a single *initial thread*, executes a single parallel region, and then continues. We show the 11 steps of this process and the terminology used to describe these steps in the OpenMP Specification.

Recall that tasks may be *tied* or untied. A tied task is always executed by the same thread. Tasks are by default tied as are any implicit tasks. If there are resources associated with a thread (such as `threadprivate` data), the tied task can depend on them always being there. An untied task, however, may be executed by different threads in the team as the task is switched from active execution to suspension. This switching occurs at well-defined *task scheduling points*. In particular, an implementation may cause any thread to suspend execution of its implicit task at a task scheduling point and switch to execute explicit tasks generated by any of the threads in the team. This is important, for example, so a thread can pause an activity that generates new explicit tasks and help the team of threads execute a backlog of tasks waiting to execute.

There are a few remaining terms we need to define before you can comfortably read the OpenMP Specifications. Parallel regions can be nested. We can create a team of threads each of which can create their own nested teams of threads. The initial thread and any threads it has created (its descendent threads) define a *contention group*. When we consider a construct, we have to understand the *binding thread set*. This is the set of threads that are impacted by the construct or that "provide the context" for the execution of a region. The binding thread set of a `critical` construct, for example, is the contention group. The binding thread set of a `task` construct, however, is the single thread that encounters the construct. We say the thread *generates* the task.

Finally, there are two additional concepts we need to understand the full range of constructs in OpenMP. In Chapter 5 we encountered a common pattern where a `parallel` directive created a team of threads for which the parallel region consisted of a single loop. That loop would have a worksharing-loop construct. As a shorthand, we combined the `parallel` and the `for` directives to create a single *combined construct*. The behavior of this construct is identical to the separate `parallel` construct followed by a `for` construct.

When we combine constructs but the semantics are different than the sequence of separate constructs, we say that the construct is a *composite construct*. A good example of a composite construct was discussed in Section 12.3. Once we move from the host device onto the target device using the `target` construct, we mapped a loop onto a GPU with the directive:

```
#pragma omp teams distribute parallel for simd
```

That directive plus the associated loop is a composite construct. Its semantics cannot be defined by a sequence of nested constructs (`teams`, `distribute`, `parallel`, `for`, `simd`).

This completes our journey through the jargon of OpenMP. We have not defined the full vocabulary of OpenMP that you will find in the Specifications. We have covered, however, the most important terms. With the terms we have covered, you should be able to pick up any others you need as you read the Specification.

## 13.3   The Structure of the OpenMP Specification

The definition of the OpenMP language began as two separate documents: one for C/C++ and a second document for Fortran. Most of us were new to the practice of writing specifications and frankly, it showed in the quality of the specifications. Over the early years of OpenMP it proved unwieldy to maintain two documents. We would find an error, for example, in OpenMP and fix it for one language but it would take us a year or longer to get the change into the specification for the other language.

OpenMP 2.5 was a major transition. Our goal was to merge the two language Specifications into a single Specification as we moved from the 2.0 Specifications to 2.5. Our intent was to add no additional functionality and just focus on how we would cover both languages in one document. We thought this would only take six months and that we would quickly wrap it up and move on to newer concepts for OpenMP 3.0.

That turned out to be quite naive. It took us over three years of difficult work to create OpenMP 2.5. We did more than just combine the Fortran and C/C++ Specifications. We carefully reconsidered both the jargon and the structure of OpenMP. The OpenMP 2.5 Specification was our sixth OpenMP Specification. By that time, we were all seasoned pros at how to construct a good specification. It was excellent and the basic structure established with OpenMP 2.5 holds to our latest, version 5.0 Specification.

We open the Specification with a detailed glossary of all the terminology used in the language. We put the glossary at the beginning to encourage readers to at least skim it and learn where to go to find specific definitions before moving into the Specification itself. We then describe the fundamental abstractions behind OpenMP, in particular the Execution and Memory models. New to the 5.0 Specification, we then define the OpenMP tools interfaces (which is not covered at all in this book).

It is with the second chapter that we move to the directives themselves. After describing their format and background details, the fundamental directive definitions are presented. Each directive follows a similar format with the following sections:

- **Summary**: A few lines that define what the directive does.

- **Syntax**: The specific syntax for a directive with separate sections for Fortran and C/C++.

- **Binding**: To which execution entities (such as threads) does the directive connect to.

- **Description**: A detailed narrative defining the behavior of the directive.

- **Events and Callbacks**: Two sections defining the tools interface.

- **Restrictions**: Rules that constrain the clauses to the directive or how it is used in your program.

- **Cross References**: References to other elements of OpenMP that impact the directive.

Depending on the directive, this basic structure may be followed by additional subsections expanding on some features of the directive. For example, for *parallel* the basic directive definition is followed by two subsections. The first section explains the rules for determining the number of threads in a team created by a parallel construct. The second section defines how the directive interacts with thread affinity issues (important for NUMA systems).

As an example, let's walk through the beginning of the parallel construct in the OpenMP 5.0 Specification. We provide a copy of the page from the Specification in Figure 13.3. This is Section 2.6 in the Specification. The name of the directive (parallel) and its status as a construct (i.e., it is a directive plus a structured-block of code) are given in the title. We then get a one-line description; that this construct creates a team of OpenMP threads. The wording is very carefully selected which is why we spent so much time discussing the terminology. In particular, the Specification indicates each thread in the team executes a region which means the code visible in the structured block immediately following the parallel directive *plus* any functions called by that code.

In lines 6 through 20 we see the syntax of the directive for the C/C++ language. The specific text of the pragma is given in line 8 with a series of optional clauses

3 ## 2.6    `parallel` Construct

**Summary**

The parallel construct creates a team of OpenMP threads that execute the region.

**Syntax**

——————————————— C / C++ ———————————————

The syntax of the **parallel** construct is as follows:

```
#pragma omp parallel [clause[ [, ] clause] ... ] new-line
    structured-block
```

where *clause* is one of the following:

> **if** (*[***parallel** :*] scalar-expression*)
>
> **num_threads** (*integer-expression*)
>
> **default** (**shared** | **none**)
>
> **private** (*list*)
>
> **firstprivate** (*list*)
>
> **shared** (*list*)
>
> **copyin** (*list*)
>
> **reduction** (*[reduction-modifier , ] reduction-identifier* : *list*)
>
> **proc_bind** (**master** | **close** | **spread**)
>
> **allocate** (*[allocator :* ] *list*)

——————————————— C / C++ ———————————————

Figure 13.3: **Parallel Construct from OpenMP 5.0** – A reproduction of the parallel construct description from the OpenMP Specification.

shown up until the newline. Since this is a construct rather than a stand-alone directive, the structured block associated with the directive is indicated on line 9.

The important feature to look up when working with the OpenMP Specification is the list of clauses that can be used with the directive. Some of these are included in the OpenMP Common Core: (*private, shared, reduction*). Each clause is either defined in the later description of the construct or elsewhere in the Specification with a specific location for where to find additional information given in the *Cross References* section.

After repeating the syntax definition for Fortran, we move on to the section on *Binding*. This section describes how the construct connects or "is bound" to execution entities such as threads and other OpenMP constructs. For example, in the parallel construct the construct is bound to the encountering thread. For the *worksharing-loop construct* and *task* constructs, the binding thread set is the current team. Since parallel regions (and hence, teams of threads) can be nested, the Specification goes on to specify that the team in question binds to the "innermost" enclosing *parallel* region. The following *Description* section provides the detailed description of the construct. This is where the details of a directive and its behavior are unambiguously described.

The OpenMP Specification can be difficult for an application programmer to read. What most of us do is we have the Examples document and the Specification side by side. Comparing the Specification to the examples for a particular directive usually is enough to figure out the meaning of the Specification. Fortunately, it gets easier with practice.

## 13.4   Closing Comments

We have now arrived at the end of the last chapter of our book. We started at the beginning assuming little or no knowledge of parallel computing. Over the course of this book, we have covered the subset of OpenMP we call the Common Core. If you followed our advice and wrote lots of code as you worked through the Common Core, you have mastered the most essential elements of OpenMP. When combined with our discussion of OpenMP "Beyond the Common Core", you truly have the bulk of OpenMP that most programmers use at your command.

This final chapter, however, embraces the one constant in modern life: change. OpenMP changes. Hardware changes. The algorithms you will need to work with change. A programmer must constantly live in "learning mode": ready to pick up new skills in the face of change. This is why we closed the book with a discussion of "Your Continuing Education in OpenMP". Books such as this one and the book *Using OpenMP – The Next Step* are powerful resources to help you get started. Over time, however, no book will be able to keep up with OpenMP. Hopefully, with the information in our closing chapter, you will be ready to go straight to the source and continue your journey into the future with OpenMP.

# Glossary

**Address space** The memory in a computer system is accessed by address. A variable is a name for an address in memory. The *address space* is the set of all addresses available to a process. In the OpenMP Common Core, we assume a shared address space is available to all the threads associated with a process.

**Amdahl's law** Amdahl's law is a simple relation which shows that the part of a program that is not parallelized limits how much faster a program might run when executed in parallel with multiple processors. If *alpha* is the serial fraction, that is, the fraction of the program that cannot run in parallel, *Amdalh's* law states that at best, a program can run $1/alpha$ times faster.

**Atomic operation** An atomic operation is an operation that can not be observed in a partial state. It is either "complete" or it has not yet "occurred". Only one thread at a time can execute an atomic operation and it can not be interrupted. Atomic operations are used to establish ordering constraints between threads (i.e., for synchronizing threads). If two or more threads use the values of shared variables to coordinate their execution (e.g., to implement a spin-lock), the only way to do so without creating a data race is to read and write those shared variables with atomic operations.

**Barrier** The *barrier* is the fundamental synchronization construct in the OpenMP Common Core. A barrier defines a point in the execution of a program where threads in a team wait until all of the threads in the team have arrived. Once all of the threads have arrived at the barrier, variables in shared memory are flushed (i.e., their values are made consistent with memory) and then threads execute statements following the barrier.

**Cache** A *cache* is a memory buffer that provides low latency access to a block of memory. A cache does *not* define a distinct address space. Informally, you can think of a cache as providing a window into the RAM memory of a system. Data is moved between memory and the cache in units of a *cache line* which corresponds to a contiguous segment of addresses in memory. A typical cache line in a modern microprocessor is 32 or 64 bytes. There are many ways to organize the caches in a system. Typically, there are a pair of caches close to each core. These are called L1D and L1I for the *L1 data cache* and the *L1 instruction cache*. They are small but run at or near the clock of the CPU. A larger but slower cache is the *unified L2 cache*. The term "unified" is used to signify that it holds both data and instructions. The hierarchy continues through multiple levels until we reach the last level cache which is a larger and slower cache shared between the cores in a multiprocessor.

**Cache coherence** In a shared memory system with caches for each processor, a single variable may exist in multiple locations across the memory hierarchy. Most of these systems are said to be cache coherent; that is, they guarantee that in a properly-synchronized, race-free program, the system maintains a single view of the memory. This means the system must keep track of the values across a cache hierarchy and update them as needed when processors read or write to shared variables.

**Cluster** Shared memory, in order to be effective, requires a significant investment in hardware to support a shared address space across processors with a variation in latencies to memory that is suitable for the parallel algorithms programmers write. At some point, as we scale the size of a parallel computer, the cost of maintaining shared memory becomes too high and unwieldy to implement. The solution is to transition to a distributed memory system where each computer in the system has its own distinct memory. Interaction between computers then happens as the exchange of discrete messages rather than through loads and stores into a shared address space. A cluster is the dominant way to build distributed memory systems. A cluster uses "Commercial Off The Shelf" (COTS) computers (nodes) with a COTS network to build large scale distributed memory computers. Software systems organize the nodes in the cluster so they appear as a single integrated system. The most important software in a cluster is the message passing software, typically based on the MPI standard.

**Concurrency** A condition of a system in which two or more execution entities are active but unordered. By "active" we mean the execution entities are executing a sequence of operations. By "unordered" we mean that we do not have global time stamps that allow us to say when operations from different execution entities are executed with respect to each other. When such ordering constraints are needed, we use *synchronization* operations.

**Construct** An OpenMP *executable directive* and the associated loop or *structured block*. It does not include any code in routines called from within the structured block. It only includes the lexical extent of the *executable directive*. The constructs in the common core are `parallel`, `task`, `single`, `target` and the worksharing-loop. OpenMP defines combined constructs which are made by merging two constructs together. The semantics of a combined construct is the same as if the two separate constructs are called successively. OpenMP also defines *composite* constructs which are constructed by merging constructs and directives but the resulting semantics might differ from what would follow from successive application of the individual constructs.

**Core** To improve aggregate performance, a processor is usually composed of smaller processors. When these processors appear at an abstract level as a distinct processing element with their own sequence of instructions, they are called a core. The CPUs in most high performance computing systems generally have multiple cores. A core often includes hardware elements to support multiple threads. This is called Simultaneous Multi-threading (SMT) or *hyperthreading*. Each hardware thread appears to the operating system as a *virtual core*. For example, a high-end CPU for a server may have 24 physical cores but SMT technology might support 2 hardware threads per core in which case the operating system would report 48 virtual cores.

**CPU** A Central Processing Unit is a general purpose processor optimized for low latencies and interactive use cases. By "general purpose" we mean that a CPU is expected to run any well-formed program. To support interactive use cases, a CPU typically has a cache hierarchy to hopefully keep frequently updated variables in memory buffers that run fast relative to the speed of the processing elements within the CPU. As a class of devices, CPUs are extremely common appearing in everything from high-end servers inside data centers to tiny chips running in a cell phone. In high performance computing systems, we informally think of a CPU as the device that occupies a *socket* in a server.

**Critical** The critical directive plus its associated structured block defines a synchronization construct that provides mutual exclusion in OpenMP. The code in the structured block can only be executed by one thread at a time. If a thread encounters the critical construct and another thread is already executing code in the construct, it will wait until that thread has completed the work defined by the construct, made any updates to memory visible to other threads, and exited the construct. In the computer science literature, this functionality is often referred to as a *critical section*.

**Data environment** The set of variables visible inside a region. This means that each construct (i.e., a directive plus its associated structured block) has its own data environment. OpenMP provides a set of clauses that define how variables move between data environments. The most common examples of these clauses are `shared`, `private`, and `firstprivate`.

**Data race** A data race occurs when: (1) two or more threads in a shared memory system issue loads and stores to overlapping address ranges, and (2) those loads and stores are not constrained to follow a well-defined order. The term "race" is used since the threads running on the different processors are "racing" to see which store lands in the shared variable. Most modern languages (including OpenMP) stipulate that a program with a data race is invalid; a compiler is not required to produce well-defined results in such cases.

**Directive** A directive is a command issued to the compiler and expressed within the source code of a program. In OpenMP, a directive is introduced with the sentinels `#pragma omp` in C/C++ and a comment statement such as `!$OMP` in Fortran. OpenMP is an explicit API so the directives tell the compiler to carry out a specific transformation to the code during compilation. OpenMP defines several types of directives. A *declarative directive* occurs among the declaration statements in a program and influences how variables are declared. An example is the `threadprivate` directive. An *executable* directive appears among the executable statements of a program and typically tells the program how to transform code during compilation to support threads. The `parallel` directive is a good example of an executable directive. A *stand-alone* directive is not associated with any declarations or blocks of code. It defines a direct action for the compiler to insert into the stream of instructions the compiler generates. The `barrier` directive is a stand-alone directive.

**DRAM** The memory in a typical computer system is exposed as Random Access Memory (RAM) which is usually supported by hardware modules implemented with Dynamic Random Access Memory chips. The term DRAM is used in this book when we want to specify a hardware element that supports the memory system.

**Environment variable** An environment variable is a mechanism to modify the environment within which a process executes. The details of how these variables are set and managed are not defined in OpenMP as they often vary from one operating system to another. Typically, each of the internal control variables (ICV) in OpenMP has an associated Environment Variable. This is used to set the default value of the ICV for an OpenMP execution. The most commonly used OpenMP environment variable is `OMP_NUM_THREADS`.

**First touch** The amount of memory that can be addressed by a system is greater than the amount of physical memory (DRAM). In response, an operating system organizes memory into pages where a page can fit into physical memory. If the system is a NUMA system with multiple NUMA domains, the performance varies widely depending on where the pages are mapped relative to the cores that access the pages. A common strategy in such systems is called first touch; i.e., a page of memory is mapped to the NUMA domain of the core that first touches the data. In practical terms, this means an OpenMP program should initialize data with the same threads that will later process the data.

**Flush** The *flush* is an operation that makes its set of shared variables consistent with memory. Note that a flush does not define a *synchronized-with* relation with other threads. It is not a synchronization operation. Flush, however, is essential in controlling data synchronization. The flush forces variables in registers or other buffers to be written to memory and it marks cache lines as "dirty" so they will be refreshed from memory on the next load. The flush operation is often called a "memory fence" in other shared address space systems.

**GPU** Graphic Processing Units were initially designed for processing graphics data. These are throughput-optimized devices. For example, if you are rendering an image, the time to compute any particular pixel is not important. The concern is the throughput; that is, the number of images per second that can be streamed through the GPU. Over time as more sophisticated rendering algorithms were developed, GPU processing pipelines became programmable. This led to GPGPU programming or General-Purpose GPU Programming. The execution model of GPGPU is SIMT. In OpenMP, the `target` and associated device directives are used to program a GPU.

**Internal Control Variable** An opaque object internal to an OpenMP implementation that manages default values, execution modes, or other behaviors for the execution of an OpenMP program. In most cases, an internal control variable (or ICV) has an associated environment variable and runtime library routines to set the variable and to get the value of the variable.

**Load balancing** A team of threads working together to execute code in-parallel completes their work when the last thread in the team has finished. Variation in when threads complete results in a subset of the threads waiting for other threads to finish; thereby incurring parallel overhead. *Load balancing* refers to techniques that adjust the work done by each thread so the team of threads finishes at about the same time. For OpenMP programmers, this often comes down to adjusting the parameters of the schedule clause on a worksharing-loop construct.

**Lock** A synchronization operation implemented in OpenMP through a lock data type and a collection of runtime library routines. These implement mutual exclusion execution through a pair of fundamental operations: *set* and *unset*. A thread sets the lock. We say that this thread *holds* the lock. If a thread tries to set a lock while another thread holds the same lock, it will wait until the thread holding the lock *unsets* the lock. Locks support mutual exclusion synchronization but in a way that is more flexible than with OpenMP directives such as `critical`.

**Memory** A subsystem in a computer that holds the values of variables. The memory is accessed through addresses hence we can describe the memory as the subsystem in a computer that supports the address space for the system. Memory is organized into a hierarchy with faster/smaller memory units (cache) near the processors and slower/larger memory devices (usually as DRAM modules) further away from the processors.

**Memory model** The full name is "memory consistency model" though we typically call it a "memory model" for short. The memory model is the set of rules that define the value returned by a read (or *load*) operation on a variable when that variable is shared between two or more threads. The model is used when reasoning about multiple threads that issue loads and stores to overlapping address ranges to assure that a program is free of any *data races*.

**MPI** The Message Passing Interface (MPI) is the dominant standard API for programming distributed memory computers. As the name implies, it defines semantics for how processes in a distributed memory system exchange messages. MPI is much more than a system for passing messages, however, and more accurately is a full-fledged system for coordinating the execution of processes including collective communication, one-sided communication, shared memory regions, and the basic constructs needed to build runtime systems for partitioned global address spaces. MPI and OpenMP have grown side by side over the years and the two have become the dominant models of high performance computing; with MPI between nodes and OpenMP on a node. This is often called the MPI/OpenMP hybrid model.

**Multicore** A CPU with multiple cores is a *multicore* CPU. While technically *multicore* is an adjective, it is often used as a noun. In some cases, we distinguish between multicore CPUs which connect cores through the memory hierarchy (cache coherent) as opposed to a many-core CPU which connects the cores through a scalable on-die network.

**Multiprocessor** A class of computer systems where multiple processors share a single address space supported by a physical shared memory system.

**Multithreading** An execution model in which a number of light-weight execution entities (threads) execute within a shared address space.

**Node** Large scale parallel computers are built by connecting multiple independent computers together over a network of some variety. We call each computer in this system a *node*. Another way to think about the term is the computer network defines a graph. The nodes in the graph are the computers in the network while the edges of the graph represent point-to-point links in the network.

**NUMA** A NonUniform Memory Architecture is a shared memory computer for which the cost to different locations in memory varies between the processors in the system. Because of the presence of caches, most computer systems available today are NUMA systems.

**Original variable** Variables appear inside the code associated with a construct. Many of these variables are declared prior to the construct and pass into the scope of the construct by default or through one of the clauses on the construct. For such variables, there is a variable of the same name that exists immediately prior to the construct. This is called the *original variable*.

**Parallelism** Multiple processors running at the same time to solve a problem are running *in parallel*. Multithreading is a specific type of parallelism where a collection of concurrent threads run on multiple processors to execute in parallel.

**Process** Operating systems organize the execution of programs in terms of a *process*. A process includes one or more threads and handles resources to support the threads. This includes a region of memory that is shared between the threads.

**Processor** A generic term to refer to any hardware element upon which threads can run. This includes CPUs, GPUs, DSPs, cores, and any other variety of processing element.

**RAM** When we consider memory in a computer system, we typically are referring to Random Access Memory (RAM). This is byte addressable memory that supports arbitrary streams of memory references (i.e., random access).

**Region** All code encountered during a specific instance of the execution of a given *construct* or of an OpenMP library routine. The region includes code from the structured block (i.e., the lexical scope of the directive) plus any code called as the threads execute the code within the construct.

**Runtime library** OpenMP provides a set of library routines callable at runtime to manage features of the implementation that cannot be addressed at compile time. Examples include the `omp_thread_num()` function which returns the ID of an individual member of a team of threads or the `omp_num_threads()` function which returns the number of threads in a team.

**SIMT** Single Instruction Multiple Thread is an execution model commonly used to understand execution of programs on Graphics Processing Units (GPUs). A space of indices is defined often by a nested set of loops. At each point in this index-space, an instance of a function called a kernel executes. Data is also organized around this index space which helps programmers reason about memory locality. The kernel instances are grouped into blocks which are queued up for execution and execute as their data is available. The goal of SIMT execution is to optimize the throughput of the system; that is, any individual kernel instance may take a long time to compute, but the aggregate collection of kernel instances complete at a high bandwidth.

**SMP** A Symmetric MultiProcessor is a shared memory computer where: (1) every processor is treated the same by the operating system, and (2) the cost of accessing any location in memory is the same for all processors.

**Speedup** A ratio between some reference run time and a comparison run time. It is important when reporting speedup data to specify the reference run time. Typically, we are interested in speedup trends as additional processors are used to execute a parallel program. In this case, the reference run time should be the best serial algorithm running on one node. When the speedup equals the number of processors, we say that the program is displaying *perfect linear speedup*.

**SPMD** Single Program Multiple Data is a fundamental design pattern of parallel programming. Each execution entity runs the same program (Single Program) but on its own set of variables (Multiple Data). The work is managed between execution entities through the ID of each entity and the number of entities running in parallel.

**Structured block** A block of one or more statements associated with certain OpenMP directives to define a construct. The statements in the structured block define a flow of execution where in normal operation of the program, execution enters at the top of the block and exits at the bottom. In the OpenMP Common Core, the only exception to the "enter at the top and exit at the bottom" rule is an exit statement to terminate execution of the program. For C/C++ the structured block is either a single statement (including a `for` statement) or a collection of statements between curly braces ({ and }). With Fortran, OpenMP defines a directive to mark the end of the structured block (e.g., `!$OMP parallel` and `!$OMP end parallel`).

**Synchronization** Operations from concurrent threads are unordered with respect to each other. This means that in general, we cannot say which operations on one thread happen-before operations on other threads. Synchronization refers to ways we can insert specific ordering constraints into the execution of concurrent threads. Specifically, a synchronization event defines a synchronized-with relation between threads. Operations before the synchronized-with relation on one thread *happen-before* operations on another thread that occur after the synchronized-with relation. When the synchronization applies to the whole team of threads, such as with the barrier and the critical construct, we call it *collective synchronization*. We can also define synchronization events between pairs of threads; that is, pairwise synchronization. When synchronization refers to the order of updates to variables in memory, it is called *data synchronization*. We also specialize the term to refer to constraints on the order of operations from multiple threads. This is called *thread synchronization*.

**Task** The term *task* is used informally to describe a distinct unit of work. In OpenMP, it refers specifically to a specific instance of executable code and its data environment. A task is an *explicit task* if it is created by an OpenMP `task` construct. A task is an *implicit* task if it is implied by a construct. For example when an OpenMP program begins execution, it is run by an initial thread which runs an implicit task known as the *initial task*. It may seem odd to define implicit tasks in OpenMP. They were added to the language to provide a consistent abstraction to be used when defining the detailed behavior of OpenMP for those implementing OpenMP systems.

**Thread** An execution entity with its own private memory (organized as a stack) and associated static memory, called *threadprivate memory*. In a modern operating system, a program executable is launched as a single process which defines an address space and a collection of resources managed by the operating system on behalf of the process. Execution of the process occurs through one or more threads which belong to the process and share the address space and any other resources associated with the process. Threads are a general concept and the term is used widely in computer science. Closely related to OpenMP threads are *pthreads* which is a standard threads interface included in the IEEE POSIX standard. Unfortunately, the term thread is used by some GPGPU programming models, which can be confusing since a GPU thread is quite different from threads in OpenMP and POSIX. This is why in the GPGPU programming model, OpenCL, the concept of thread was dropped and instead, the more generic term *work-item* was used.

**Thread affinity** A multiprocessor system is optimized to maximize the performance across many simultaneous processes. Given the large number of processes running on a system at any given time and the fact that they may be in various states of the activity or waiting on system resources, the most effective way to maintain good aggregate performance is for the operating system to freely migrate the threads between the processors in the system. While this is an effective strategy for servicing the needs of a collection of largely independent processes, this can be a terrible strategy when you are interested in the performance of a single process (as is usually the case with an OpenMP program). The solution is to enable thread affinity; that is, to tell the operating system to turn off thread migration and then bind threads to specific processors.

**UMA** A Uniform Memory Architecture is a computer system where the cost function for any memory access is the same for all processors in the system. An ideal SMP computer is an UMA system.

**Worksharing** A type of construct in OpenMP. A worksharing construct specifies that the team of threads will work together to carry out the work defined by the region associated with the construct. The work is divided among the threads in the team as opposed to having each thread redundantly execute the code in the region (as is done, for example, with the parallel construct). The worksharing-loop is the most commonly used worksharing construct in OpenMP.

# References

[1] Gene M. Amdahl. Validity of the single processor approach to achieving large scale computing capabilities. *Proceedings of the Spring Joint Computer Conference*, pages 483–485, 1967.

[2] Bronis R. de Supinski, Thomas R. W. Scogland, Alejandro Duran, Michael Klemm, Sergi Mateo Bellido, Stephen L. Olivier, Christian Terboven, and Timothy G. Mattson. The Ongoing Evolution of OpenMP. *Proceedings of the IEEE*, 106(11):2004–2019, 2018.

[3] Robert H. Dennard, Fritz Gaensslen, Hwa-Nien Yu, Leo Rideout, Ernest Bassous, and Andre LeBlanc. Design of ion-implanted mosfet's with very small physical dimensions. *IEEE Journal of Solid State Circuits*, SC-9(5), 1974.

[4] Intel. *Intrinsics Guide*. Available at https://software.intel.com/sites/landingpage/IntrinsicsGuide/.

[5] Hartmut Kaiser, Bryce Adelstein Lelbach, Thomas Heller, Agustín Bergé, Mikael Simberg, John Biddiscombe, and Christopher. Hpx v1.2.1: The c++ standards library for parallelism and concurrency. 2019.

[6] Leslie Lamport. Time, Clocks, and the Ordering of Events in a Distributed System. *Communications of the ACM*, 21(7):558–565, 1978.

[7] Leslie Lamport. Turing Lecture: The Computer Science of Concurrency: The early years. *Communications of the ACM*, 58(6):71–76, 2015.

[8] Tim Mattson, Rob van der Wijngaart, and Michael Frumkin. Programming Intel's 80 core terascale processor. *Proceedings of SC08: the International Conference for HPC, Networking, Storage, and Analysis*, 2008.

[9]  Timothy G. Mattson, Beverly A. Sanders, and Berna L. Massingil. *Patterns for Parallel Programming*. Addison-Wesley, Boston, MA, 2005.

[10] John D. McCalpin. Stream: Sustainable memory bandwidth in high performance computers. Technical report, University of Virginia, Charlottesville, Virginia, 1991-2007. A continually updated technical report. http://www.cs.virginia.edu/stream/.

[11] Aaftab Munshi, Ben Gaster, Timothy G. Mattson, James Fung, and Dan Ginsburg. *OpenCL Programming Guide*. Addison-Wesley, Boston, MA, 2011.

[12] Feng Niu, Benjamin Recht, Christopher Re, and Stephen J. Wright. Hogwild!: A lock-free approach to parallelizing stochastic gradient descent. In *Proceedings of the 24th International Conference on Neural Information Processing Systems*, NIPS'11, pages 693–701, USA, 2011. Curran Associates Inc.

[13] Ruud van der Pas, Eric Stotzer, and Christian Terboven. *Using OpenMP – The Next Step*. The MIT Press, Cambridge, MA, 2017.

[14] John von Neumann. First draft of a report on the edvac. *"Contract No. W-670-ORD-4926 between U.S. Army Ordnance Dept. and the Univ. of Pennsylvania"*, 1945.

[15] Anthony Williams. *C++ Concurrency in Action*. Manning, Shelter Island, NY, 2012.

[16] Samuel Williams, Andrew Waterman, and David Patterson. Roofline: an insightful visual performance model for multicore architectures. *Communications of the ACM*, 52(4):65–76, 2009.

# Subject Index

## Scientific and Engineering Computation

William Gropp and Ewing Lusk, editors; Janusz Kowalik, founding editor

*Using MPI: Portable Parallel Programming with the Message-Passing Interface, third edition,* William Gropp, Ewing Lusk, and Anthony Skjellum, 2015

*Using Advanced MPI: Beyond the Basics,* Pavan Balaji, William Gropp, Torsten Hoefler, Rajeev Thakur, and Ewing Lusk, 2015

*Scientific Programming and Computer Architecture,* Divakar Viswanath, 2017

*Cloud Computing for Science and Engineering,* Ian Foster and Dennis B. Gannon, 2017

*Using OpenMP—The Next Step : Affinity, Accelerators, Tasking and SIMD,* Ruud van der Pas, Eric Stotzer, and Christian Terboven, 2017